André de Lima

Study of the quality of parts obtained in an optimized drilling process

André de Lima

Study of the quality of parts obtained in an optimized drilling process

Surface integrity in the drilling process

ScienciaScripts

Imprint

Cover image: www.ingimage.com

This book is a translation from the original published under ISBN 978-3-330-77237-3.

Publisher:
Sciencia Scripts
is a trademark of
Dodo Books Indian Ocean Ltd. and OmniScriptum S.R.L publishing group

120 High Road, East Finchley, London, N2 9ED, United Kingdom
Str. Armeneasca 28/1, office 1, Chisinau MD-2012, Republic of Moldova, Europe
Managing Directors: Ieva Konstantinova, Victoria Ursu
info@omniscriptum.com

Printed at: see last page
ISBN: 978-620-8-64258-7

SUMMARY

1 INTRODUCTION

With the intense demand for productive agility resulting mainly from the process of globalization experienced since the 1990s, research centers, industries and other segments related to the development of productive activities have come under strong pressure to take drastic action regarding the time spent on manufacturing products.

Since the mid-1970s, the evolution of machining in the areas of turning and milling has been remarkable, given the amount of research carried out on these subjects. The combination of modern modular tools with new cutting edge geometries, more wear-resistant materials and new coatings has made it possible to increase cutting speeds to levels unimaginable until the early 80s. On the other hand, the latest technology machine tools, which can work at very high cutting speeds and axis travel, allow machining cycles to be shortened more and more (KALHOFER, 1997).

Competitiveness has driven companies to study the various possibilities for reducing costs, especially in sectors directly linked to production, investing in technological innovations in machining processes (very high cutting speed - HSC) and in obtaining new materials (tool and product). (MULLER, 2000).

Along with these innovations, economic and environmental demands have suggested the use of dry cutting or minimum quantity cutting fluid (MQL) processes, atomized by compressed air, to replace emulsified cutting fluid (BRAGA, 2001).

On the operational side, there has been a lot of focus on increasing the order of magnitude of machining parameters. However, these increases have not always been closely monitored by specialists in the field; more often than not, they have been the result of empirical decisions, made by professionals without the necessary qualifications and who, more often than not, do not have the necessary knowledge of the demands to which the parts produced will be subjected in the field (MULLER, 2000).

However, one area of machining has not kept up with this progress: solid rotary tools, mainly drills and taps, which CRUZ, 1978, already pointed out were lagging behind; this has only been remedied in recent years with the development of new specific technologies (MULLER, 2000).

Among the machining processes that have undergone this type of investigation, the drilling process, the setting for this research, has undergone major transformations aimed at improving and optimizing the process. These transformations are the result of research aimed at balancing knowledge and technological advances in the drilling process with those of other conventional machining processes with a defined cutting edge geometry (BRAGA ET AL, 2001; MULLER, 2000).

In view of the importance of drilling in manufacturing processes, there has been a great deal of research aimed at improving processes (generally through new materials and technologies for manufacturing drills, eliminating cutting fluids or optimizing machining conditions) (SALES ET AL, 1998), but little attention is paid to the macro and micro determinants of quality (MmdQ) of machined products, and more interest is taken in optimized results (life, time, etc.). MIRANDA, 2000, is a clear example of this type of approach, as he sought to "explore the limits of use of solid carbide drills, with a view to analyzing the economic and productivity conditions of these limit conditions", with no concern for the results in terms of MmdQ.

This scenario has sparked the interest of some research groups in studying more closely the relationships between the improvements introduced in the search for optimization of the drilling process (increases in cutting speeds, elimination of lubrication and cooling) and the MmdQ. VIEIRA ET AL, 1998, have already pointed out that shape and/or dimensional deviations and the integrity of the subsurface layer can suffer significant alterations due to the use of optimized conditions in limit situations.

In this context, this work aims to analyze, based on results obtained in a drilling process without cutting fluid and with optimized cutting parameters, the behavior of MmdQ as a function of this machining condition. It also aims to initiate a discussion on the occurrence of possible failures that arise when parts are used and which may originate from the alteration of these determinants due to thermal and mechanical stresses arising from the use of tools in extreme conditions.

1.1. Objectives and Structure of the Work

The main objective of this work is to use MmdQ concepts to evaluate the viability of the cutting parameter optimization strategy in a drilling operation without cutting fluid. The parameters of this analysis will be some of the MmdQ measured on specimens obtained in tests previously carried out with a view to optimizing the process and using the tool up to its life limit.

2 LITERATURE REVIEW

This chapter presents a set of concepts about the drilling process that are necessary for understanding the proposal; it provides notions about machining optimization and process optimization methodologies; and it presents a brief overview of the main technological changes that have been introduced into the drilling process. These changes are treated as an evolution not only of the process, but of the entire scenario surrounding the drilling process (tools, machining conditions, etc.);

2.1. DRILLING PROCESS

The history of drilling machines is certainly one of the longest and oldest among machine tools. It is a history that has been marked by a visible supplanting of this equipment in the face of dynamic demands from the industrial world (DINIZ ET Al, 1999, CRUZ, 1978).

Throughout this history, drills have undergone many changes in an attempt to at least partially meet the market's momentary needs. This path of change has been marked by the emergence of increasingly robust and technologically advanced drills. However, even with these innovations, drilling machines tend to be replaced in many operations by machining centers, due to their mechanical limitations and lack of flexibility, among other aspects.

Contrary to what has been happening with drilling machines, the drilling process is currently experiencing a totally different reality: there is a concentration of studies aimed at a market demand that incessantly seeks a reduction in costs followed by an increase in productivity (DINIZ ET Al, 1999).

The drilling process is one of the most widely used in the manufacturing industry in general, and is widely exploited in the automotive sector. This can easily be seen from the fact that most parts in any type of industry have at least one "hole", and that only a very small number of these parts come with the hole ready or almost ready from the process of obtaining the raw part (DINIZ ET Al, 1999).

This is the case with sintered parts which have part of their holes requiring only dimensional and shape calibration (VIEIRA ET AL, 1998). In general, parts have to be drilled in full or have their holes enlarged through the drilling process, which makes the study aimed at optimizing this process very important.

To explain the share of drilling operations in the metal-mechanics industry, we use as a basis the survey carried out by CARDOSO, 2000, in a large auto parts supplier. For the manufacture of an 8-cylinder diesel engine cylinder head, 53% of the operations carried out are drilling (Figure 1). Other examples could be taken, but it is not the scope of this work to identify the volume of drilling operations in the industry; we just want to show that these operations can be significant, justifying the attention that has been paid to them.

Table 1 shows the results obtained with strict controls of the drilling process, still within the framework of the analysis developed by CARDOSO, 2000. It can be seen that increases in the machining parameters (cutting speed and feed) can effectively optimize the process and reduce the costs of the operation (useful life increases by up to 1500% and the cost of the operation can be reduced by more than 90%).

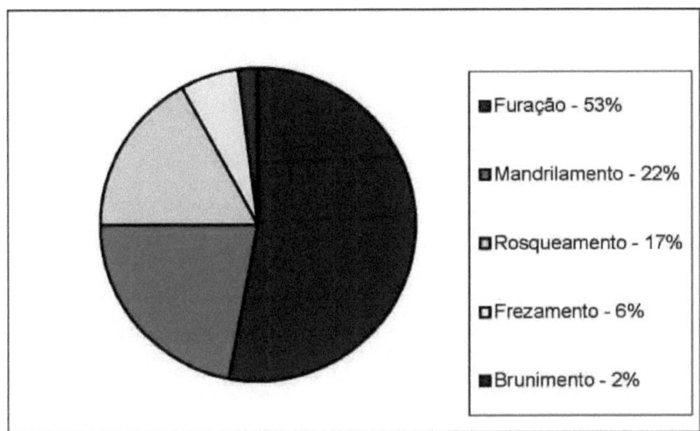

FIGURE 1 - MACHINING operations INVOLVED IN THE MANUFACTURE OF AUTOMOTIVE ENGINE CYLINDER HEADS.

SOURCE: CARDoSo (2000).

TABLE 1 - RESULTS OBTAINED BY CHANGING CUTTING PARAMETERS IN A Drilling PROCESS

Parameters	**Before**	**After**	**Gain %**
Cutting speed	80 m/min	IOOnVmm	20%
Progress	234 mm/min	732 mnVmin 500 mnVmin	213% 114%
Operating Time	0.6 min.	0.25 min.	58.3 %
LJtil Life	30 pcs	500 pcs. 180 pcs	1567% 500%
Cost of Operation	R$ 3.70 / pc	R$0.21 / pc	94%

SOURCE: CARDoso (2000)

In the drilling process, the cutting parameters are often limited by the maximum speed of the machine tool spindles, their displacements and the rigidity of the machine-tool-device-workpiece system. The drilling process lags behind other machining processes for a number of reasons, the main one being the diameter of the holes. A 10 mm diameter drill bit, to be used at speeds compatible with carbide (Vc around 200 m/min), would need a speed of around 6400 rpm, something far beyond the limits of conventional drills. For this reason, the number of CNC machining centers being used for drilling has grown (DINIZ ET Al, 1999).

Another reason for this lag was already pointed out by CRUZ, 1978: "twist drills have not had the same level of development as other machining tools, either in terms of tool geometry or the materials used".

Given the importance of drilling in machining processes, new materials and processes for manufacturing workpieces and drills have emerged, producing tools with a wide range of hardness and toughness (SALES, 1998). However, the most commonly used drill bits are still made of high-speed steel. This is partly due to a lack of information on the part of those involved in decision-making within companies.

This is one of the main reasons for tool breakages observed in cases where the industry tries to apply new machining technologies (HSC, drilling without cutting fluid, etc.) in order to optimize the process. As the incidence of breakages is one of the main reasons identified by CARDOSO, 2000, for the occurrence of failures and stoppages in the production process, innovations end up being discarded and the process remains unchanged, maintaining the condition of technological backwardness.

On the other hand, there are research groups working to overcome the limits of the tool/machine/part combination. They often find new resources to break through the barriers to process optimization. In the case of drilling, new tools have been developed, with very complex surfaces, edges and cutting angles, seeking to extract something more from the operations than reducing production time and costs (SOTO, 2000), which will be discussed below.

2.2. PROCESS EVOLUTION

Drilling tools are among the oldest tools used by man. Drilling tools first appeared in the 1770s, when bits were used to drill wood. Even then, the geometric shape of the bits was very similar to the shape of the twist drills used until then. From 1822 onwards, the drill bit was also used for drilling metal (FERRARESI, 1977).

However, with the great leap forward made by industrialization in the mid-19th century, there was a demand for a more efficient drilling tool that would allow higher working speeds. This sparked a desire in the sectors involved to quantify, or rather to compare one operation with another in a quantitative way, thus being able to argue and propose improvements, which possibly gave rise to the concept of machinability (WADA, 1997).

The machinability of a metal can be defined as a technological quantity that expresses, by means of a comparative numerical value, a set of machining properties of the metal, in relation to another as a standard. These properties include measurable quantities inherent in the metal machining process, which can be listed as tool life, cutting effort, productivity and chip characteristics (WEINGAERTNER and SCHROETER, 1991; FERRARESI, 1977). To this list can be added the determinants of product quality, factors that gained importance at the end of the 20th century (VIEIRA ET AL, 1998).

Tool development has advanced so rapidly that the problems of thermal stress, which occur when drilling without coolant, can be overcome. New geometries allow hot chips to be removed from the hole without heating up the tool. Machine tools are being developed so that their reactions are almost inert to temperature, allowing chips to be removed without the use of cutting fluids (KAMMERMEIER, 2001).

2.3. Drilling Tools

The tools discussed in this chapter are, in general, drills, especially twist drills, as these are the tools that are the object of study in this work.

To better understand this tool, figure 2 shows some basic forms of construction, design and some of the parts that make up a twist drill, according to DINIZ, 1999:

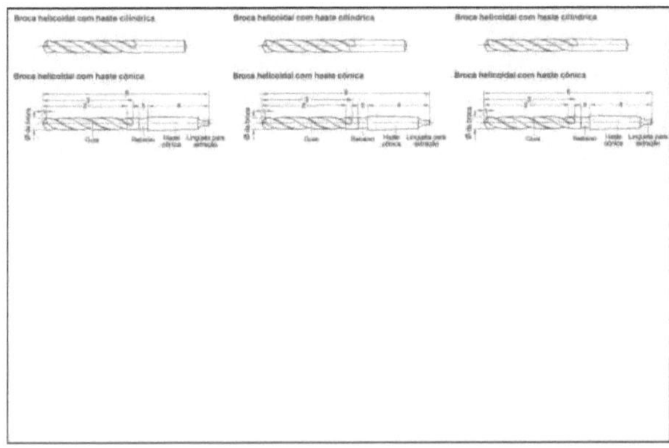

Figura 2- Helical Drill Parts With Cylindrical Stem And With Conical Stem.

SOURCE: DENIZ ET AL, 1999.

***S* Shank** - used to fix the drill bit to the machine. For drills with a small diameter (up to 15mm), cylindrical shank drills are generally used and are fixed to the machines using chucks. For drills with larger diameters, conical shanks are preferred, which are clamped to the machines using a Morse cone, which allows for greater clamping force;

***J* Guides** - these are the external surfaces of a drill bit which appear as two regions (one on each cutting edge) which have a diameter greater than the diameter of the drill bit walls. They have two basic functions - the first, as the name implies, is to guide the drill bit into the hole; the second is to prevent the entire outer wall of the drill bit from rubbing against the walls of the hole, thus reducing the effort required for drilling;

J **Helical grooves** - these are the tool's exit surfaces. The helix angle of standard drills can be 28° for general-purpose drills (N-type drills), 15° for drills intended for machining materials with short chips (H-type drills) and 40° for drills intended for machining materials with long chips and/or soft materials (W-type drills). The length of the helical groove can also vary depending on the diameter of the drill bit and the depth of the hole to be machined;

J **Diameter (0)** - is measured between the two drill guides, the drill diameter usually has a tolerance of h8, it is of fundamental importance to the process as it gives the machined part the dimensional characteristics desired in the project;

J **Core** - inner part of the drill bit with an approximate diameter of 0.16 0. It guarantees the drill bit's rigidity; e.

J **Cutting edges** - on a twist drill, the two main cutting edges do not meet at a point, but there is a third edge connecting them. This third edge is called the transverse cutting edge. The angle formed by the main cutting edges is called the cutting edge angle (DINIZ ET AL, 1999).

In most machining tools, the mechanical and thermal loads on the cutting edge are quite significant, especially in the drilling process where the cutting speed varies along the radial direction (Figure 3), allowing the generation of a false cutting edge which makes chip removal more difficult. As a result, the heat generated by high cutting speeds is removed less, with a consequent increase in the temperature of the drill cutting edge, which can compromise the life of the drill (BRAGA ET AL, 2001).

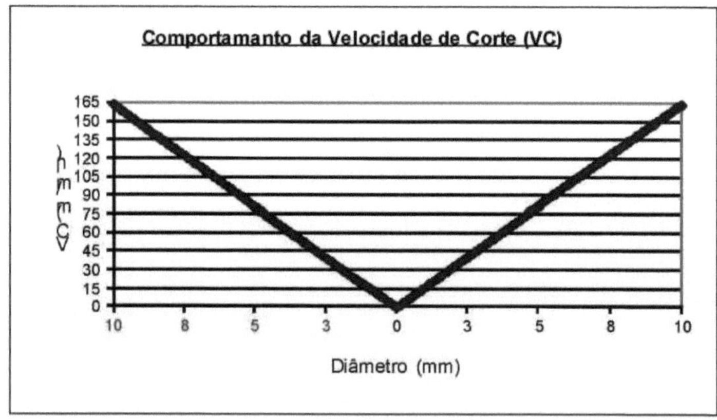

Figura 3- EXAMPLE OF THE VARIATION OF THE CUTTING SPEED ALONG THE RADIAL DIRECTION OF THE CUTTING EDGES (U =10MMEN=5188 RPM).

Figure 4 attempts to summarize some of the main problems encountered in drilling without cutting fluid, which have been solved over time as a result of scientific and technological innovations in the areas of drill development (MULLER, APUD BRAGA, 2001).

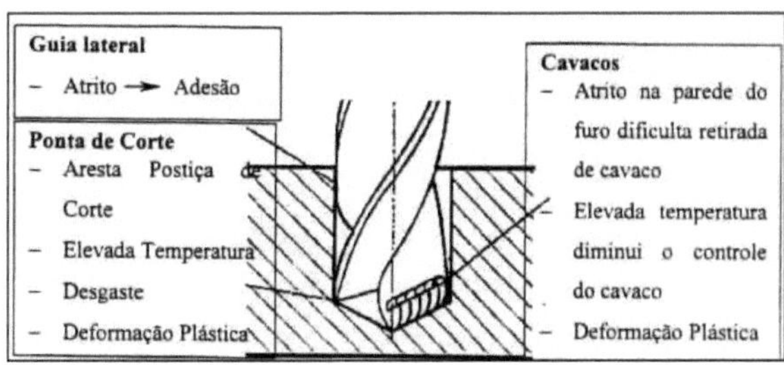

Figura 4- MAIN PROBLEMS ENCOUNTERED IN DRY HOLE DRILLING

SOURCE: MULLER, apUD BRAGA, 2001.

With regard to the friction between tool and chip, which causes material adhesion and the consequent formation of a false edge, drills have been given coatings such as TiN-Al; with regard to the friction between the chip and the hole wall, drills with special profiles have been developed; other innovations will be discussed later.

When producing a hole, all the requirements must be taken into account: diameters, depths, tolerances, production and the tools available for the job. A fundamental rule is that twist drills always work under the same condition: the cutting speed varies from zero, at the center of the drill, to the maximum value at the periphery of the tool (end of the main edges of cut) (YUHARA, 1982) as shown in FIGURE 3.

With greater emphasis on the complexity of the drilling process, FIGURE 5 shows the cutting speed ranges for drilling and tapping machining processes using taps (rotary tools), for working with HSC machining technology (MULLER, 2000; SOTO, 2000). In the case of drilling, the cutting speed range indicates that, in comparison with the data in FIGURE 3, in the same machining operation the main cutting edges work at cutting speeds ranging from those considered conventional to speeds typical of HSC. This is a type of behavior that does not occur in other types of machining processes (turning, milling, for example).

Figura 5 - *CUTTING SPEED RANGES FOR HSC MACHINING IN DRILLING AND TAPPING.*

SOURCE: (MuLLER, 2000).

Among the developments in the drilling process, some concepts stand out, such as the development of machine tools with more than one axis and new tooling concepts. A clear example of this are machine tools that perform the multiple drilling process, or rather, machines with several spindles (polymandrel). In this case, an attempt is made to make up for the time spent at low cutting speeds by simultaneously drilling all or at least most of the holes in a part, thus reducing the time spent producing the part. This technology marked a phase in the evolution of the drilling process that preceded the current phase, when attention began to focus on new geometries and materials for producing drills (DEGARMO, 1979; FREIRE, 1978). An example of a machine tool with several chucks (polymandrel) can be seen in Figure 6.

Figura 6 *EXAMPLE OF A POLIMANDRIL HEAD DRILL.*

Machines with polymandrel heads, however, had the problem of being machines with limited use, given their characteristics of inflexible configuration. Today, machining centers are being used to perform drilling operations when it is necessary to combine high cutting speeds with operational flexibility.

Still within the innovation process, figure 7 shows the incorporation of several innovative concepts into a single tool. The use of new "S" cross-sectional edge geometries and increasingly tough carbide substrates can be highlighted, with the aim of alleviating the problem of cutting speed in the center of the drill; there is also the issue of reinforced PVD coatings with or without internal cooling (SOTO, 2000).

Figura 7 MODERN ALTERNATIVES IN CARBIDE DRILLS.

SOURCE: (SOTO, 2000).

2.4. ELIMINATION OF CUTTING FLUID IN DRILLING

The main functions of cooling/lubricating fluids are: to reduce heat generation by reducing friction; to eliminate the influence of the increase in temperature of the tool/part/machine assembly; and to transport/remove the chips generated during the process; protection of the part and machine against atmospheric corrosion (KLOCKE & EISENBLATTER, 1997).

Although manufacturers of machining fluids come up with various nomenclatures to define the types of fluid, chemically, cutting fluids are divided into two large groups:

S Integral fluids or pure oils free of water; e.

S Water-based fluids.

In the machining process, cutting fluids have two properties: as lubricants or coolants.

Water/oil systems were soon introduced and offer the advantage of the cooling provided by water and the lubrication provided by oil derivatives. To them are added extreme pressure additives, microbiological growth control additives and anti-corrosion protectants (NOVASKI and RIOS, 2001).

The correct selection of the most suitable cutting fluid for a given machining operation is not a simple task. It requires a careful assessment of the various aspects involved. When applied correctly, cutting fluids can increase productivity and reduce costs, making it possible to use higher cutting speeds, feeds and depths of cut. In addition, they increase tool life, improve the surface finish of the machined part and reduce the cutting power consumed (RODRIGUES and ABRÂO, 1999).

On the other hand, the costs associated with the use of cutting fluid correspond to between 14 and 16% of the cost of machining (KONIG, 1998). MULLER, 2000, tries to show how this correspondence occurs and indicates at what level each of these components can be reduced (figure

8).

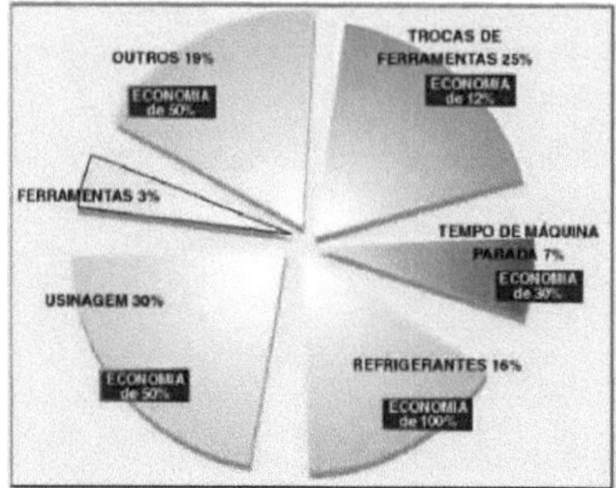

FIGURE 8 - The IMPACT OF COOLANTS ON TOTAL MACHINING COSTS.

SOURCE: (MuLLER, 2000).

Cutting fluid costs can be saved by up to 100% in the case of drilling without the use of cutting fluids. Other costs can also be reduced by working to reduce tool wear (by up to 12%) or optimizing the process (by up to 50%).

As a result, in addition to working with optimized parameters, work began without the use of cutting fluids, with the aim of eliminating waste and reducing production costs, given the price of lubricants and the equipment involved in lubrication (for example, the care taken to avoid the development of bacteria in storage tanks) (NAKAGAWA, 2000).

The cost of using cutting fluids has risen sharply, mainly due to their high consumption and constant disposal, which is in direct conflict with environmental preservation policies; it is necessary to find systems that are compatible with the operation, the user and environmental preservation.

Major environmental concerns have been raised in recent years as a result of market demands (ISO14000 and the "green seal"). Work has been carried out on eliminating the use of products that generate chemical residues (chlorine, sulphur and phosphorus) that are harmful to man and nature, including the lubricating and cooling fluids used in machining processes (especially drilling) (SALES, 1998).

In line with global environmental control trends, which are increasingly stringent and focused on environmentally friendly manufacturing (NAKAGAWA, 2000), these requirements are contained in highly important standards for companies aiming to be recognized as "World Class Enterprises" or aiming to partner with them (CARDOSO, 2000).

In the last decade, a lot of research has been done to minimize the use of coolants and/or lubricants in metalworking production. In addition to the operational costs of production and environmental issues, there are legal requirements regarding human health issues due to the contact of the coolant with the skin, breathing and/or ingestion of pollutants derived from them (BRAGA, 2001; HEISEL & LUTZ, 1998; KALHOFER, 1997; SALES ET AL, 1998, LIMA ET AL, 2001, KLOCKE ET AL., 1997).

Nowadays, when dealing with manufacturing systems, the main focus is on increasing productivity while reducing costs. In machining operations, the use of cutting fluids suitable for the operation is an alternative for increasing productivity that should be included in process analysis.

When drilling at high cutting speeds (HSC) it is necessary to use high pressure when injecting the fluid, as this is the only way to reach the contact zone (workpiece/drill); but the fluid must be resistant to evaporation, as in these situations it is more likely to evaporate due to the pressure and temperature conditions to which it is subjected. In addition, the fact that when fluid is injected at high pressure, it becomes a kind of barrier to chip extraction cannot be overlooked.

When machining without cutting fluid, friction and adhesion between tool and workpiece increase due to the increased thermal load involved in the process. This leads to significant increases in tool wear, the formation of craters and elongated chips which cause problems, especially in drilling processes. These difficulties can currently be minimized with the use of coated tools with modified geometry (KLOCKE ET AL, 1997).

2.5. Cutting Fluid Disposal and Drilling Tools

With the trend towards reducing the use of cutting fluid in machining operations, there is a concern to develop tools with special chip-breaking geometries for better control of chip discharge and also in the tool designs themselves to prevent premature wear of the cutting edges when operating without cutting fluid. The choice of material used to make the tool is just as important as the geometry of the tool chosen, as it determines the hardness characteristics of the tool under high temperature conditions and its resistance to wear (SCANDIFFIO, 2000).

It is important to note that the substrate of carbide tools used for cutting without lubrication must be produced from ultra-fine powders, with grain sizes smaller than 0.3 □ m and high heat resistance. These characteristics make it possible to obtain sharper cutting edges, which generates less heat compared to traditional carbide tools. These ultra-fine powders provide tools with approximately 60% to 70% more resistance than those obtained with conventional carbide powders (grain size of approximately 2.5 microns) (CSELLE, APUD SCANDIFFIO, 2000).

To complete the process of obtaining a good tool for machining conditions without the cutting fluid, titanium carbide and/or aluminium oxide, titanium nitride and titanium carbonitride coatings are added to the characteristics mentioned above, which are essential to overcome the effects of lubrication deficiency in the cutting region. According to DUNLAP, 1997, these coatings reduce friction and adhesion, acting as a kind of "solid lubricant"; in addition, the thermal load on the substrate is reduced due to the low thermal conductivity of the coating layer. The reduction in heat dissipation by the tool changes the flow of heat between the tool and the chip, which causes the chip to dissipate more heat than normal, thus increasing cutting time due to reduced wear on the cutting edges (SCANDIFFIO, 2000).

3. MACRO AND MICRO DETERMINANTS OF QUALITY (MMDQ)

Macro and Micro Determinants of Quality (MmdQ) are the set of characteristics and occurrences generated in parts after machining. Generally speaking, parts subjected to any machining process have quality determining factors, which are part of the parts' identity.

The MmdQs chosen for analyzing the specimens, the main objects of study in this work, are concentrated on shape, dimensional and location errors, which are macro determinants, and on the surface and subsurface characteristics of the holes (stress, micro-hardness and affected layer), which are micro determinants.

The interest in this set of MmdQ is justified by the following reasons:

S Compliance with the general rules of geometric tolerances is a basic requirement for the quality of holes. It is important to note that the diameter, perpendicularity and location of the hole are parameters of fundamental importance for acceptable part quality (GRIFFITH, 1994; PUNCOCHAR, 1996; AGOSTINHO, 1977);

S The existence of mechanical and thermal effects that affect the surface layer in machining processes, as shown in Figure 9 (JACOBUS ET Al, 2000).

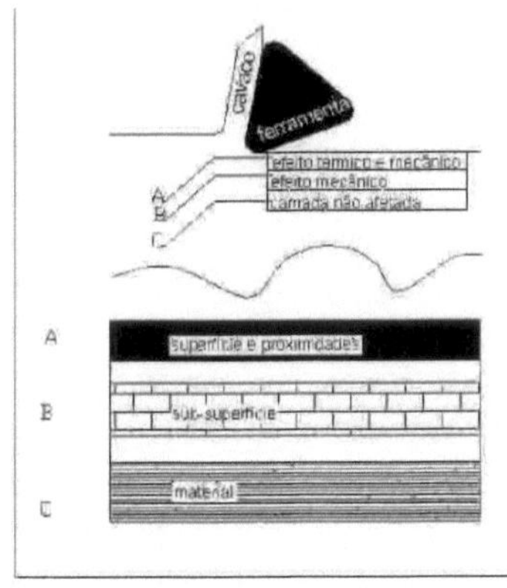

FIGURE 9 - SCHEMATIC OF THERMAL AND MECHANICAL EFFECTS ON MACHINED PARTS.

SOURCE: (JACoBUs ET AL, 2000).

Figure 9 shows the moment of material removal in a machining operation. The workpiece has been subdivided into three portions: A, B and C, which differ according to the type of stress they are exposed to. In section A, the surface of the workpiece and its surroundings, there is a kind of combination of thermal and mechanical stresses; in section B, the subsurface of the workpiece, there is a possible mechanical stress; in section C, the core of the material, there is a region unaffected by stresses. It is in portions A and B that the main changes in the micro determinants of quality occur.

The following are some concepts and discussions about the MmdQs of interest, highlighting their influence on the final result of the drilling process.

3.1. Dimensional and location deviations

The process of obtaining a part involves all the time spent, from conception, design and execution. These phases include a whole history of dimensional definitions, manufacturing tolerances and a whole range of information which, in most cases, is insufficient to accurately state which quality standard the parts fall into once they are finished (GRIFFITH, 1994; PUNCOCHAR, 1996).

By comparing the actual part manufactured with the ideal part specified by the project and depicted in the drawing, it can be said that they differ significantly. The degree of difference between the actual part and the designed part characterizes manufacturing precision, and is determined by the quality of the machining. This is usually foreseen in the design phase, and is specified in advance in the form of so-called dimensional and location tolerances (AGOSTINHO et al. 1977; MATEOS, 1974; PUNCOCHAR, 1996; GRIFFITH, 1994).

These deviations arise during the execution of the part by the machine tools and are caused, in most cases, by the inaccuracy of the processes in terms of machine rigidity, tool wear, machining parameters adopted for the process and countless other factors that directly influence the final quality of a machined part (BALAKSHIN, 1983; GRIFFITH, 1994). In the specific case of drilling, the loss or reduction in the rate of chip removal from the main cutting edge.

In the design of a particular part, in addition to the fitting and manufacturing tolerances, the so-called geometric tolerances or geometric deviations must also be taken into account, in order to obtain the best possible functional quality (NOVASKI, 1998).

In all finished parts, obtained with or without chip removal, you can see not only the deviations allowed by the corresponding tolerances, but also other types of deviations. NOVASKI (1998) defines them as macro-geometric (when related to shape and location) and micro-geometric (when related to surface roughness).

Among the macro-geometric deviations that occur in a machining process, the most important in this group are deviations in dimension, parallelism, perpendicularism, circular shape (circularity), cylindrical shape (cylindricity), angle and location (MATEOS, 1974).

In this work, the macro-geometric deviations most focused on are deviations in size (diameter), perpendicularity and location (position of the holes). These deviations were chosen based on the characteristics of the tests, which are similar to a roughing process or the initial phase of drilling the hole, which precedes a hole calibration stage carried out by a tool that is more sensitive to wear, which is why diameter control is of fundamental importance if the tool is not to wear excessively, which is just one of the reasons for choosing these deviations as the object of analysis.

3.1.1. Dimension deviations - Diameter

Dimension deviations are related to problems that occur in manufacturing processes: tool wear, deformation of the MFDP assembly (machine - tool - part - device), thermal expansion, among others (LIRANI, 1985).

Among the deviations that occur in the drilling process, dimensional variation in diameter is the most frequent, and is usually associated with deviations in roundness and cylindricity. Considering that a hole must be used as a functional element for the passage of another element (pin, shaft, screw, etc.), diameter variation can be considered critical in drilled parts (FARAGO & CURTIS, 1994).

3.1.2. Perpendicularity deviations

It is considered that the right angle is a creation of the human mind, because in nature its presence is not deliberate. In man-made products, however, whether in architecture, commerce or industry, in civil engineering, the right angle has been sought after as the result of the mutual perpendicularity of lines and surfaces, and has been present since ancient times. This is justified by the varied and intense application of the right angle in many cases. They can be related to technical convenience and feasibility, but also to less distinct concepts; they could reflect the human mind, which searches for an expression of regularity or a mathematically definable definition (FARAGO & CURTIS, 1994).

The perpendicularity deviation is used to characterize functional and/or constructive elements that need to have an angular relationship of 90° between their parts, such as an interface between two surfaces. It must guarantee the axial positioning of holes in relation to the surface of a part, so that these holes can have the function of a fastening or orientation element (guide hole) (GRIFFITH, 1994; PUNCOCHAR, 1996).

For the specific case of bores, the deviation from perpendicularity is checked through a tolerance zone contained between two parallel planes of infinite rotation that characterize a cylinder. These planes must be referenced to a surface of the workpiece, generally the hole's entrance surface, as shown in figure 10 (GRIFFITH, 1994).

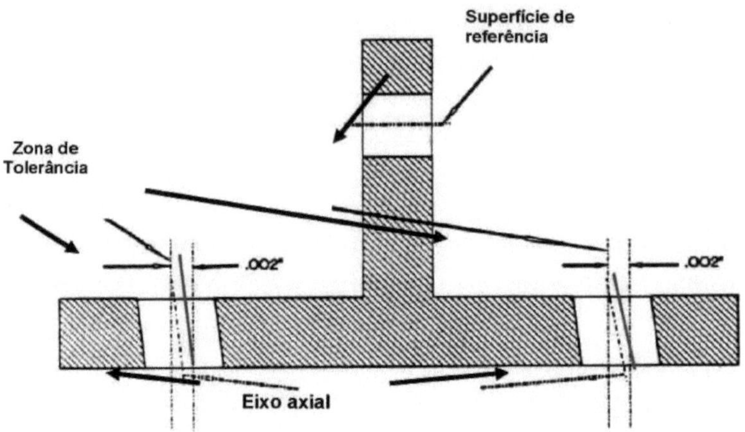

FIGURE 10 - EXAMPLE OF PERPENDICULARITY DEVIATION FOR HOLES

SOURCE: (GRIFFITH, 1994).

The presence of a right angle in a geometric part is of great operational importance, as it plays a fundamental role in product development and engineering. Consequently, perpendicularity is a condition that must often be measured in parts or parts of parts (FARAGO & CURTIS, 1994).

3.1.3. LOCATION DEVIATION

Location deviation is a type of location deviation and can be defined as the difference between an edge, surface or element of the part in question and the theoretical location prescribed by the design of the part. This theoretical location is determined by angle and distance tolerances based on a reference system previously adopted on the part itself (edges or surfaces); this reference system is based on Cartesian or polar coordinates (AGOSTINHO, 1977),

When machining fixing holes for covers, which have to be fixed to housings using screws and guide pins, this specification is essential to ensure assembly interchangeability. The location deviation will be limited by the location tolerance (TL), which can be presented in three ways:

S Point location tolerance;

S Line location tolerance;

S Plane location tolerance.

For this work, we are interested in the location tolerance of a point, which is determined by a spherical surface or a circle with a diameter of TL, the center of which is determined by the nominal measurements, as shown in figure 11. This tolerance is widely used in sheet metal drilling.

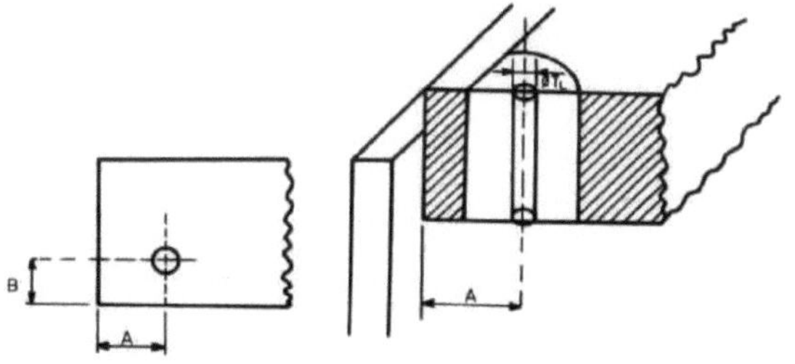

FIGURE 11-EXAMPLE OF POINT LOCATION TOLERANCE

SOURCE: AGOSTINHO, 1977.

The degree of complexity of the location tolerance increases as the quantity and quality of the information gathered in the analysis process increases. This complexity also depends on the shape of the part or object to be inspected Figure 12.

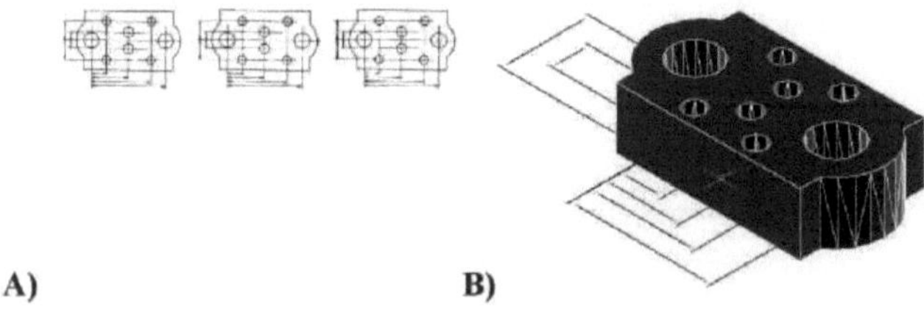

A) B)

FIGURE 12 - EXAMPLES OF LOCATION TOLERANCE

Figure 12 shows two different types of part, although the composition and arrangement of the holes are similar, with only the thickness varying. In this case, if we take the symmetry of the upper and lower faces of part B as the basis for analysis, a variable will be inserted in relation to the analysis of part A, which would be the difference in position of the axial axis of the hole that could interfere with the coordinates of one face to the other. One of the causes of this variation, for example, could be a deviation from perpendicularity.

3.2. SURFACE INTEGRITY

Together with the need for shape and location, surface integrity decisively determines or exerts a certain control on the behavior of workpieces, such as the propagation of cracks (DAMASCENO, 1993).

Damage to the part's surface integrity can accelerate the part's fatigue process, alter its resistance to abrasion and corrosion, or even cause cracks to appear and propagate (VIEIRA ET AL, 1998).

The impairment of the surface integrity of machined parts can be defined by surface alterations, which are the result of the demands of the machining processes used to obtain these parts (COSTA, APUD DAMASCENO, 1993). Among the types of surface alterations associated with chip removal work, the following can be mentioned:

J Micro hardness;

J Surface roughness (which is not of interest in this work);

J Surface and subsurface alterations (CAC);

J Residual stresses.

The factors of interest for this work are described in more detail below.

3.2.1. Heat Affected Layer (HAC)

The analysis of the CAC, also known as the *HAZ - Heat affect zone*, has been gaining ground in the machining sector. Well known in welding processes, the HAZ is proving to be a very useful and effective tool for ascertaining quality within one of the micro quality determinants (VIEIRA ET AL, 1998).

In machining processes, much of the mechanical energy involved is transformed into thermal energy, i.e. heat; it is customary to speak of "heat generated in the process" to refer to this transformation of energy. This heat has three possible sources (DEGARMO, 1997):

J The formation of the chip itself is a process in which the main sources of thermal energy are the elastic and plastic deformations of the material and the separation of the material from the workpiece;

J Friction between tool and workpiece;

J The friction between the chip and the tool.

The heat generated in the cutting region can be dissipated through the tool, the workpiece, the cutting fluid and dissipation in the air; the first two forms are undesirable because they accelerate the tool wear process and can promote the formation of CCS, since temperatures in the chip-forming region can reach up to 1000°C (SNOEYS APUD ABRÂO, 1991).

With the rise in temperature and subsequent cooling of the workpiece, a favorable condition is created for the appearance of layers of pure martensite (□'). According to SHAW, 1994, although there is no long exposure time to high temperatures as is the case with heat treatments, the phase change occurs due to the plastic deformations to which the material is subjected at the same time as the temperature increase during chip formation.

These deformations, combined with heating, promote a coordinated change of atoms in a short space of time and result in the formation of a thermally affected layer (TAC). To better understand the formation of the TAC, it is interesting to look at figure 9, which illustrates what happens to the structure of the material in certain portions of the surface towards the center of the piece.

In addition to the drilling process, other machining processes have adopted the formation of CAC as a parameter for analyzing damage to the surface layer, such as GUNTER SPUR and HOK TIO, 1991, LIMA ET AL, 2001 and LIMA & VIEIRA, 2002, who used this analysis for the grinding process.

3.2.2. MICROHARDNESS

The mechanical property called hardness is widely used in the specification of materials, in mechanical and metallurgical studies and research, and in the comparison of different materials. However, the physical concept of hardness does not have the same meaning for all areas that deal with this property. This divergent conceptualization depends on the experience and field of activity of each area studying the subject (DIETER, 1976). For example:

S For metalworking, hardness means resistance to permanent plastic deformation;

S For mechanical engineering, hardness is defined as the resistance to penetration of one harder material into another;

S For machining, hardness provides a measure of the metal's resistance to cutting;

In grinding, the hardness of grinding wheels is associated with the resistance of the wheel alloys to retaining the abrasive grains (VIEIRA, 1996).

It is therefore not possible to find a single definition of hardness that encompasses all of these concepts, not least because for each of these meanings of hardness, there are one or more indicated ways of measuring it.

From this point of view, the hardness test can be divided into three main types, which depend on the way in which the test is conducted (SOUZA, 1982):

1. By penetrating one harder material into another. Hardness is measured according to the load applied and the depth of the impression left by the penetration;

2. By shock between two materials, one of which is of known hardness. Hardness is associated with the mechanical energy involved in the measurement process;

3. By scratching, in which a material of known hardness (usually diamond) is placed in contact with the material of interest and a sliding action is made, which leaves an impression (scratch). The hardness is associated with the dimensions of this impression.

The first type mentioned was the one used for the analysis developed in this work, more precisely with the use of Vickers microhardness measurements.

The principle of measurement by the Vickers method is as follows: the penetrator is a diamond tip, in the shape of a pyramid with a square base, with an angle of 136° between the opposing faces, on which the load for the measurement is applied. As the penetrator is a diamond, it is practically deformation-free, and as all the impressions are similar to each other, the Vickers (HV) method is independent of the load, i.e. the hardness value obtained is the same whatever the load used for homogeneous materials.

For this hardness measurement method, the load varies between 1.00Kgf and 120.00Kgf for conventional Vickers hardness and between 0.01Kgf and 1.00Kgf for microhardness analysis, according to ASTM E384-99. The change in load is necessary to obtain a regular impression, without elastic deformation of the material, and of a size compatible with the measurement of its dimensions on the machine's display. The shape of the impression is a regular diamond as shown in figure 13, and by averaging L of its diagonals squared, the area of the identified surface is obtained; dividing the value of the applied load by this area gives the Vickers hardness reading given in Kgf/mm^2, as shown in equation 1 (SOUZA, 1982).

$$HV = 1{,}8544 \times P / L^2 \quad \text{..(1)},$$

Where: P→ applied load, in Kgf

L→ average of the print diagonals

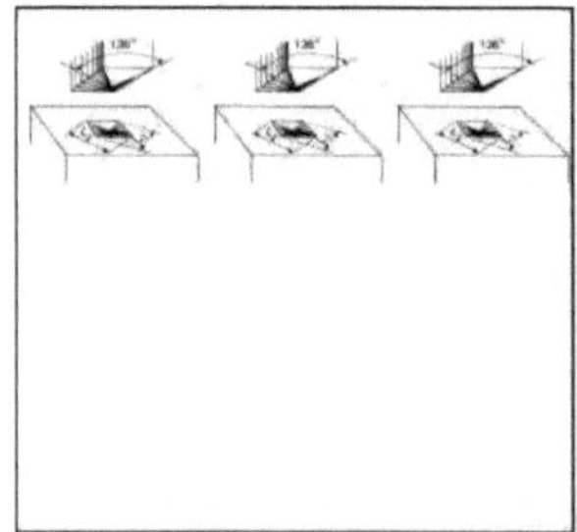

FIGURE 13 - SIEVE AND VICKERS PRINTOUT

SOURCE: (SOUZA, 1982).

The area of the impression must be accurately measured using a light microscope attached to the machine in order to determine the diagonals with great precision (around 1 □m). The preload is applied lightly to the flat surface of the sample by means of a track and is maintained for about 10 seconds; the full load is applied for a further 15 seconds, with a tolerance of 2 seconds (ASTM E384-99).

The main advantages of the Vickers method are (MARIN, APUD SOUZA, 1982):

S Scale continues;

S Extremely small imprints that do not render the part unusable compared to the Brinell method;

S High measurement accuracy;

S Zero deformation of the penetrator;

S There is only one hardness scale;

S Application for the full range of hardnesses found in various materials;

S Can be applied to any thickness of material, allowing surface hardness to be measured.

Vickers hardness is widely used for research, studies and more specifically for determining the depth of surface protection layers and the depth of decarburization in steels (SOUZA, 1982), hence the adoption of this method for carrying out analyses of Heat Affected Layers (HAC), because many of the applications of Vickers hardness are geared towards microhardness testing in view of the fact that it produces a microscopic impression on the material, provided by loads of less than 1 kg (DEGARMO, 1979; SOUZA, 1982).

3.2.3. RESIDUAL STRESSES

Stress analyses show that component failure is generally related to surface defects that appear at the manufacturing stage (forming, raw or sintered machining, etc.) or as a result of a subsequent chip removal process. Depending on the manufacturing process applied and the direction of machining in relation to the direction of the load applied, it can generate significant differences in the strength of the components analyzed (GUNTER SPUR and HOK TIO, 1991).

In general, in terms of machining, there has been a great deal of interest in recent years in finding ways to optimize the production process. As a result, it is necessary to create situations that favor the performance of the product, manufacturing and use trilogy. It is therefore necessary to use suitable equipment, materials and tools to improve the surface integrity of the machined part (DAMASCENO 1993).

Residual surface tension is of fundamental importance in the behavior of the fatigue resistance of materials, i.e. when the tensile tensions are compressive, they are beneficial, improving the fatigue resistance of the material, but when they are tensile, depending on their magnitude, they are harmful, contributing to a decline in the resistance of the material (DAMASCENO 1993; VIEIRA ET AL, 1998).

During the formation of the chip at the machining point, high restoring forces are produced, inducing a state of residual stress on the surface and subsurface of the workpiece. An increase in the feed rate raises the residual surface tension, which can be a decrease in the tension that was already compressive or an increase in the case of traction (GUNTER SPUR and HOK TIO, 1991; DAMASCENO, 1993).

4. WORK PROPOSAL

Considering the set of MmdQ described in the previous chapter, it can be seen that in drilling operations, especially those carried out under optimized conditions, it is possible for problems to occur in the part as a result of the process.

Because it is generally carried out without cutting fluid and under severe machining conditions, optimized drilling is involved with high temperatures in the cutting region and high machining forces (Figure 14).

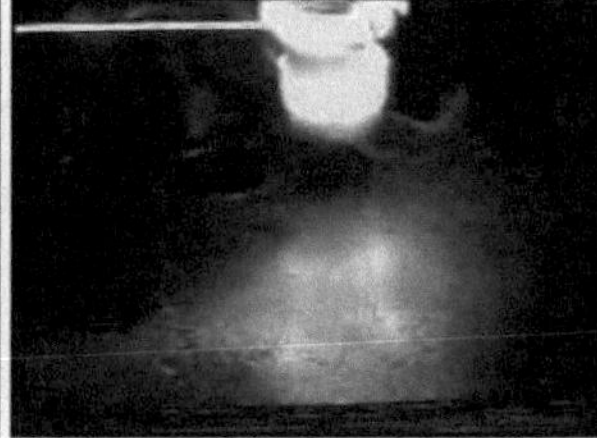

FIGURE 14 - IMAGES OF THE DRILLING OF THE SPECIMEN

Figure 14 shows two images taken at different times during the test: the image on the left shows a situation where the test was halfway through, i.e. approximately two hundred and fifty holes had already been drilled. Intense heating can be seen in the drill bit and in the vicinity of the hole. The other image in figure 14 (right) shows the behavior of the tool and chips at the end of the drill bit's life, i.e. at the end of the test. There is widespread overheating between the tool/part/chip combination, a scenario that is conducive to the emergence of MmdQ problems.

This study proposes the use of MmdQ concepts to evaluate the feasibility of the process optimization strategy used.

To this end, we intend to carry out a set of MmdQ analyses in order to identify, through these parameters, the moments at which the use of the optimized operation starts to introduce quality problems and non-conformities into the process due to excessive tool wear. It also increases costs by increasing the geometric wear of the drill tip and the coatings used, making it impossible to reuse them through regrinding, which means buying new drills.

It is important to note that due to the fact that there are already results from optimized operations, no new drilling tests were carried out, as these results would be repeated in their entirety. The optimization strategies adopted to carry out these tests were the same as those used previously by MIRANDA ET AL, 2001, given that this strategy is one of the most widely used.

5. MATERIALS AND METHODS

The equipment and materials needed to carry out the work, as well as the methods used to measure and analyze the results, are described below.

4.1. OPTIMIZING THE DRILLING PROCESS

In order to carry out the analysis proposed in this work, specimens machined under optimized machining conditions were used. These tests were carried out by MIRANDA ET AL, 2001, and aimed to obtain the limits of use of solid carbide drills in optimized drilling operations. These tests were not concerned with aspects other than the productivity of the limit conditions.

The specimens were made from ABNT 4340 alloy steel, and the holes were machined in two types of specimens. The first type (CP1) was an alloy plate measuring 430 x 310 x 32 mm, in which most of the holes were machined, simulating production in order to cause the drill bit to wear out.

After machining a series of 16 holes with a diameter of 10 mm, a hole was drilled in the second type of specimen (CP2 - with dimensions of 41 x 41 x 32 mm), which was attached to a dynamometer. During the machining of the holes made in this CP2 specimen, MIRANDA ET AL, 2001, monitored the twisting moment (Mt) and the feed force (Ff). The drawings of the two types of specimens can be seen in Figures 15 and 16, respectively.

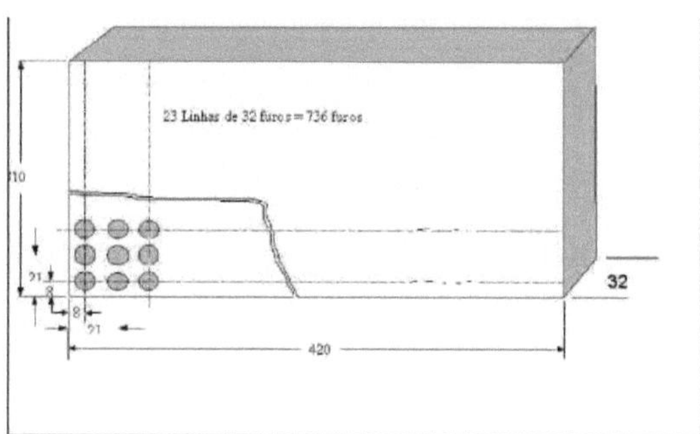

FIGURE 15 - SPECIMEN CP1 420X310X32 MM

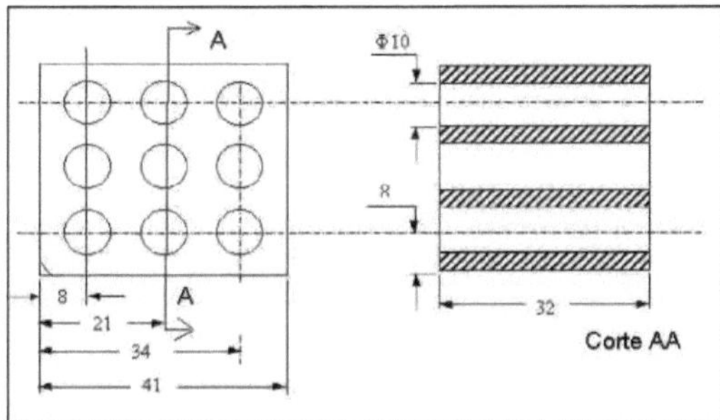

FIGURE 16 - SPECIMEN CP2 41X41X32 MM

Figures 15 and 16 show the layout of the holes and the dimensions of the specimens involved in the tests; it is possible to see important details such as the distance between holes, the thickness of the plates, among others.

The tools used, selected from the TITEX PLUS catalog, were solid carbide P40 drills, with a nominal diameter of 10mm, type ALPHA 2 right-hand version, in accordance with DIN 6537K and DIN 6535HA for the shank, with a TINAL FUTURA coating (model A3265 TFL). The use indicated in the catalog is for machining steels and cast materials up to 1300 N/mm^2, especially suitable for machining carbon steels without cooling. For machining ABNT 4340 alloy steel, the catalog recommendation is to use soluble oil or oil as the cutting fluid.

The tool clamping system used was the Hidro - Grip Coromant Capto, cone model C5-390.272-40/040 and chuck model C5-391.CGA-20/074, both from SANDVIK. The eccentricity (chatter for the drill and clamping system), verified by sampling, was on average around 0.010 mm.

The hole depth of 32 mm ensured that the ratio between the length of the hole and the diameter of the drill bit was 3.2. Figure 17 shows the eccentricity of the main cutting edge of the drill bit.

FIGURE 17 - CHECKING EXCENTRICITY

In this work, the analysis of the specimens involved two different stages: the assessment of the macro quality determinant behavior (perpendicularity, location and diameter) and the assessment of the micro quality determinant behavior (micro hardness, surface layer integrity and sub-surface of the drilled part).

4.2. MMDQ ANALYSIS

This section attempts to describe, in great detail, how the analysis of the macro and micro determinants of the Quality of the piece was carried out throughout the work.

4.2.1. DIMENSIONAL AND LOCATION ANALYSIS

In order to evaluate the behavior of the macro quality determinants, the specimens were measured on a Coordinate Measuring Machine (CMM), with the aim of identifying the dimension error (diameter) found in the drilled holes, verifying the perpendicularity of the machine tool's working axis in relation to the reference plane adopted for the measurement, and also verifying the occurrence of hole positioning errors due to drill bit wear. The CMM used was a Starrett, model DCC - RGDC 2828-24.

The measurement procedure adopted can be better explained by breaking it down into stages:

S The test specimen is positioned on the machine stand (machine support plane) supported at three points by three specific supports with millimetre height adjustments (figure 18):

S In order to check that the CP is statically level with respect to the machine support plane, the CP is zeroed using dial gauges, using regions close to the three support points as a base;

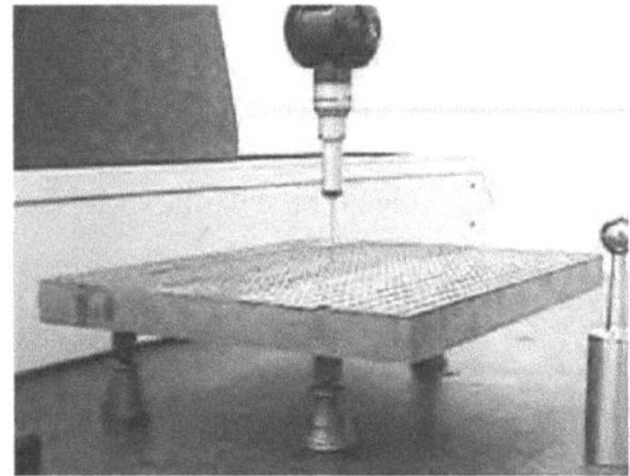

FIGURE 18 - CP POSITIONING EXAMPLE

J The probe is zeroed and the machine is referenced, routine procedures that precede a measurement sequence;

J CNC program for the automatic execution of the measurements was then drawn up;

J A reference plane is defined for the machine by means of 12 points along the upper surface of the specimen, which are randomly distributed with the sole concern of mapping all levels of the surface;

J Two reference lines are defined which are perpendicular to each other and are used to orient the machine in relation to the layout of the holes in the CP;

J The measurement sequence adopted must be the same as that adopted for drilling the holes, i.e. the first hole machined must also be the first hole measured, and so on. This strategy is adopted in order to reproduce as closely as possible the conditions suffered by the PC throughout the tests;

J For automated measurements, the program generates a routine of repetitions that are carried out using the lines generated in the initial phase of the program as a guide;

J If one of the borehole columns of the test specimen (the last column) is not completed at the end of the drill bit's life, it is necessary to take two measurements: - one for the completed hole columns and one only for the last column (figure 19);

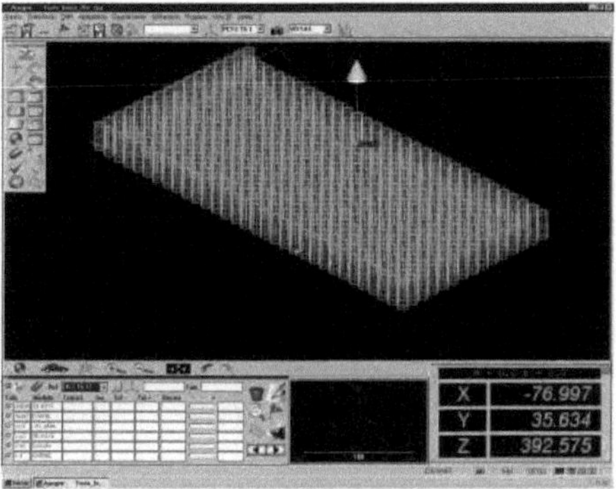

FIGURE 19 - EXAMPLE OF MEASUREMENTS FOR COMPLETED COLUMNS

S Once the program has been defined, the hole measurements begin;

The data obtained from the measurements is exported as a text file for later processing in Excel. The aim was to create behavior curves for the measured quantities (diameter, location and perpendicularity) over the life of the drill bit. This data export stage was time-consuming, as the file loses its formatting during the data transition, so all the data is exported in the form of a single huge column, which has to be restructured again.

This is the only way to arrive at the values of the measured quantities in a way that makes it possible to manipulate these values outside the software environment of the Coordinate Measuring Machine.

4.2.2. CCS ANALYSIS

The aim of this section is to demonstrate how CCS analysis should be carried out. It is important to note that the basis for this analysis is the use of image acquisition and analysis resources. To this end, a Hitachi monochrome CCD camera coupled to a metallographic microscope was used. The image captured by the camera was analyzed using *Global-Lab Image View software,* specifically designed for image analysis. The acquisition and analysis procedure is described below:

S The start of this analysis is marked by the cutting of the specimens, which must be carried out in two stages: in the first stage, an industrial hydraulic band saw is used as a tool, with which the specimen is divided into four pieces, due to its dimensions not being compatible with metallographic "Cut-Off" equipment. At this stage, care is taken to ensure that the cut is not made close to the areas that will be subjected to future analysis, given that this machine is not suitable for metallographic purposes and could therefore cause damage to the surface of the piece that could compromise the results;

S After separating the specimen into 4 pieces, it was then possible to cut the specimen in equipment directed towards the materials area (Cut-Off), where the specimen was cut into 16 holes, in order to extract a small sample from the cross-section of these holes. The number of holes for cutting the sample was stipulated on the basis of the sampling frequency used by MIRANDA ET AL, 2001, to measure the moment and shear forces;

After obtaining the samples, which have been previously identified, they are then inlaid. This must be done in bakelite using a specific machine for this purpose (inlayer), to avoid the formation of interfaces between the sample and the inlay, also known as the "edge effect";

S This is followed by all the metallographic preparation, which involves grinding, polishing and etching (nital, 2%);

S With the sample polished and attacked, and with the appropriate identifications, the analysis phase of the CCS throughout the life of the drill begins. This is effectively the beginning of the CCS measurement and behavioral analysis phase; however, what has been described so far is of great importance, as *poor* sample preparation can lead to unreliable analysis, thus giving unwanted uncertainty to the results;

S The analysis phase of CCS involves acquiring images of the samples, magnified in the microscope between 200X and 800X, and calibrating the measurement software; for each magnification value of the images, a calibration with the respective magnification must be carried out, which makes it possible to measure the sample using the images obtained and analyze the results.

Software calibration consists of converting a measurement known to the software - the number of pixels in an image - into a common unit of measurement (millimeters, meters, inches, etc....). It is done as follows:

S In order to calibrate the image analyzer, it is necessary to use a graduated scale used in optical microscopy to calibrate eyepieces and other devices; this scale is a 1mm micro-scale printed on a mirror, with centesimal subdivisions (0.01mm) (figure 20);

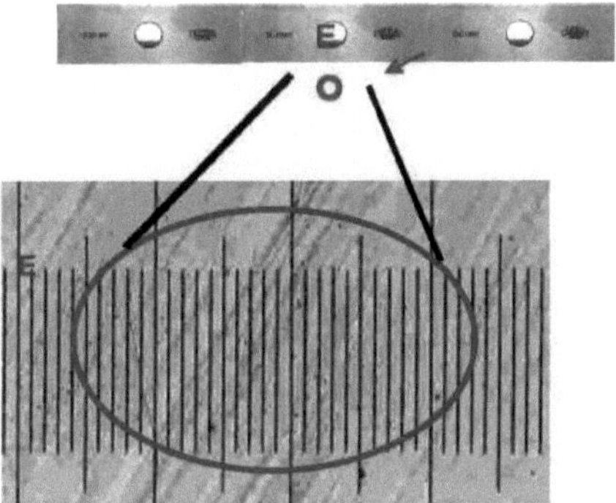

Figura 20 - *MICRO-SCALE IMAGE HIGHLIGHTING DETAIL E (MIRROR).*

S The calibration process begins by capturing images of the microscale in the vertical and horizontal positions, at all the magnifications offered by the microscope to which the camera is attached. Once these images are available, the operator is asked to determine eight known points, which are marked on the microscale images taken at the same magnification and in different directions (vertical and horizontal). The points are determined on the images of the micro-scales opened sequentially on the screen and with the calibration tool activated; thus, the points determined on one image are superimposed on the other image when it is opened (figure 21).

S The determination of these points follows a Cartesian logic with coordinates on the X and Y axes; thus, for each point determined on the micro-scale gradations, the operator provides the coordinates that determine the point. To do this, the calibration tool has four measurement windows per point, two for each axis; the first two windows contain the software's reference measurements, i.e. the pixel location of the point on the screen; in the other two windows, the operator enters the measurement value read from the microscale, i.e. the coordinates of the point (figure 22). This procedure is repeated for the eight points determined within the two images, and must be repeated for all the images at their respective magnifications (figure 23);

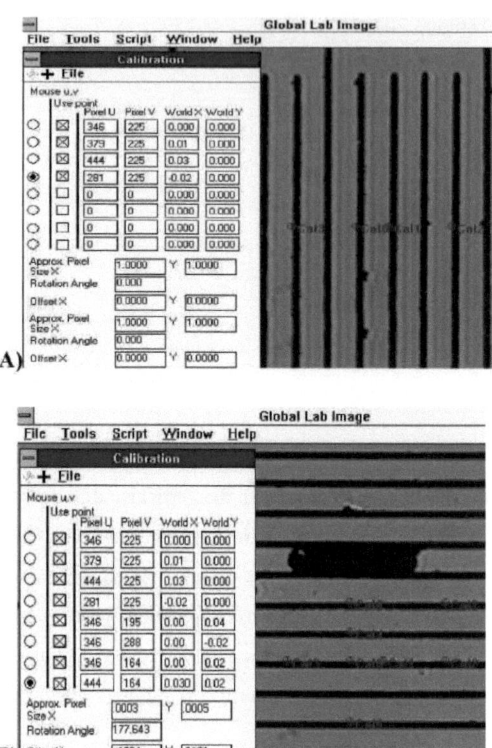

Figura 21 - *A) Partial calibration image, horizontal position. (X axis); B) Total calibration image, vertical position (Y axis), with X axis values superimposed.*

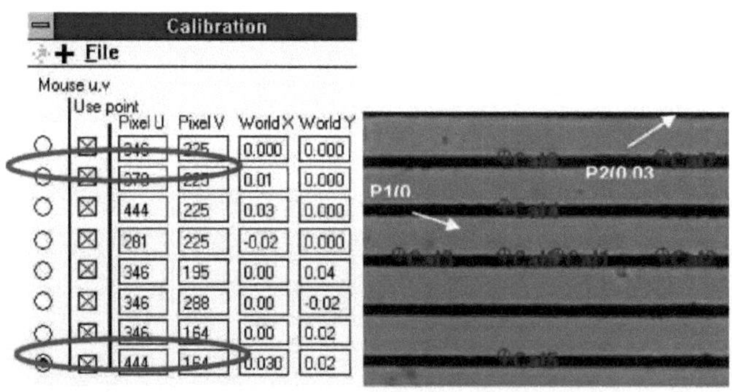

Figura 22 *Images of the calibration tool's measurement windows, and in detail the start and end points.*

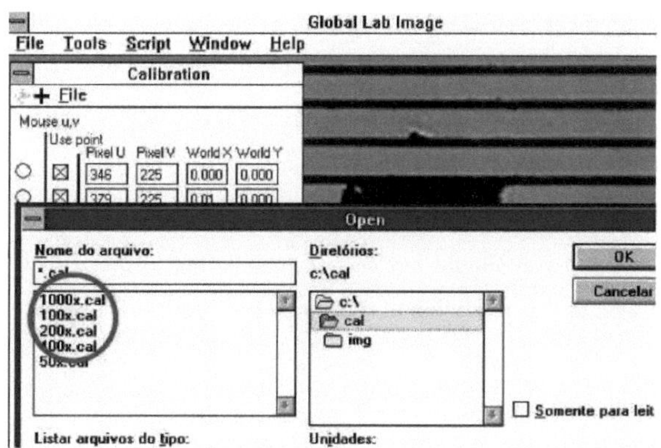

Figura 23 - DEMONSTRATION OF THE CALIBRATION FILES GENERATED TO MEET THE AMPLIFICATIONS OF THE MICROSCOPE USED.

S Once these eight points have been determined, the data is compiled; the "compute" command is given in the calibration module, which automatically performs the conversions from pixel to mm (and vice versa), and generates a file identifying the calibration performed (file extension ".Cal"). When activating any of the software's measurement modules, it is essential that the calibration file is activated at the same magnification as the image so that measurements can be taken based on the calibrated measurement values. These measurements can be given in different units of measurement (meters, millimeters, centimeters, inches and so on), depending on the gradation of the scale used.

It is important to note that the quality of the measurements is directly related to the accuracy with which the software is calibrated. Calibration must be carried out in at least two directions so that the software can make the necessary angle compensations and measurements. For the analysis in this paper, calibrations were carried out in both horizontal and vertical directions.

After calibration, it is possible to take measurements and analyze the images, which are described below:

1. Initially, the image is acquired in a region of interest in the sample, which will be analyzed and/or measured.

2. Measurement

During the image acquisition phase, certain precautions must be taken to avoid compromising the measurement results:

S Images should be acquired in regions of the sample that best represent the characteristics found in each sample. Images should also be taken of regions of the sample that differ from the whole so that a more in-depth analysis can be made of behavioral discontinuities, such as the appearance of cracks and stains. For good image acquisition, equipment such as lenses, lamps and filters must be clean and free of marks; the surface of the sample to be analyzed must be exactly parallel to the microscope table, in order to avoid problems with focusing the image; adequate lighting, well directed at the sample, is also essential;

S The captured image should be saved in image-oriented file extensions that guarantee a good resolution and do not compromise the results; you should not choose to save the images in file extensions that imply a reduction in file size, as this reduction in size could compromise the quality of the image;

S It is advisable to acquire the images under the same conditions of magnification and brightness as the calibration.

The measurement process must be carried out within the following parameters:

S Always use the correct calibration, so when acquiring the image and naming the image file, the magnification at which the image was taken should be made as clear as possible;

J Once the image magnification issue has been resolved, before opening the measurement tool, load the calibration tool with the magnification equivalent to that of the image;

J Open the measurement tool and arrange the tool menus on the screen in such a way as to allow maximum visualization of the sample under analysis;

J In view of the fact that the measurements are generated using two points selected manually on the computer screen, and taking into account possible errors in the positioning of the points selected by the operator, the measurement process should be repeated more than once and an average value adopted between the measurements taken;

J In order to reduce these possible positioning errors, the operator can use the "zoom" tool, where it is possible to focus exactly on the region desired for measurement.

In order to better visualize a measurement process through image analysis, figure 24 shows a measurement process aimed at calibrating the calibration: the measurement object becomes the scale used for the calibration obtained with the respective magnifications. This figure also shows the relationship between the measurement tool and the activated calibration tool, and in detail the area measured and the value obtained.

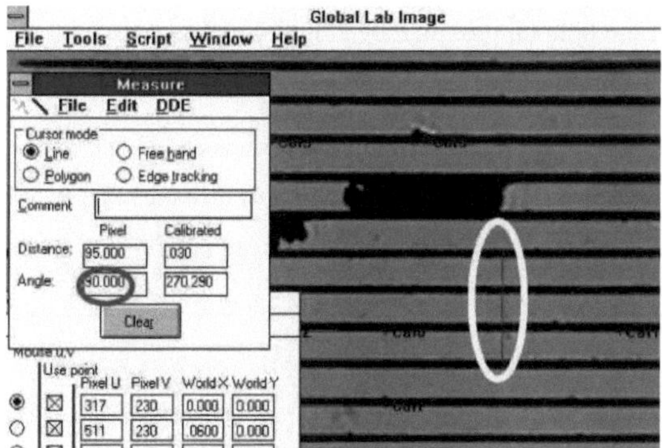

Figura 24 - *IMAGE OF A MEASUREMENT PROCESS*

5.2.3. RESIDUAL STRESS ANALYSIS

This section describes the procedure for analyzing the specimens in order to evaluate the behavior of the surface tension of the workpiece over the life of the drill bit.

As the materials laboratory at FEMP/UNIMEP does not have any equipment for measuring surface tension, the first step was to find a laboratory in the area that could provide this equipment for data collection. The analysis was then carried out at a company in the Campinas region, which provided an X-ray deflection stress measuring machine, the Digital Fastress "Residual Stress Analyzer" model: 1500 SN: 1565-98, manufactured by "Metro Design line USA".

Once this difficulty had been overcome, the measurement procedure adopted was as follows:

J A region of the specimen that has not been subjected to the effects of machining is measured in order to obtain a reference value of the stress state in which the material of the specimen is before being subjected to the drilling process;

S The measuring process begins in the opposite direction to the drilling of the holes. This is a common practice in this type of measurement. The fact that this measurement strategy is adopted can be explained by the considerable amount of time taken for each measurement, as following the reverse path leads more quickly to sudden changes in the measured values;

Starting from the last hole, the measurement is made within a predetermined interval of holes. In the case of this work, an interval of 20 holes (an arbitrary choice) was used until the measurements obtained stabilized, after which the interval of holes from one measurement to the next was increased;

S For each hole measured, the stresses are checked in three different positions (Figure 25): ***A)*** surface of the part close to the hole (approximately 0.03mm from the hole wall) as shown in the image to the left of figure 24; ***B)*** internal position of the hole close to the surface of the part (approximately 5mm deep); ***C)*** internal position close to the end of the hole (approximately 27mm deep) as shown in the image to the right of figure 25.

S These measurements are taken for two sets of holes whose position corresponds to the end of the test sequence, i.e. when the drill stops drilling holes in the main specimen and drills a hole in the secondary specimen where the feed force and torsional moment measurements are taken;

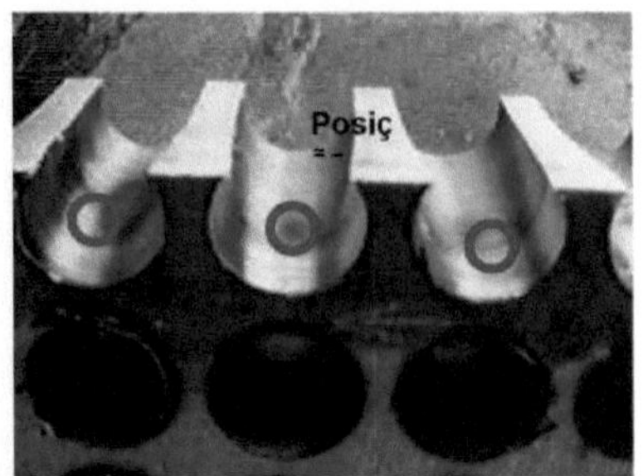

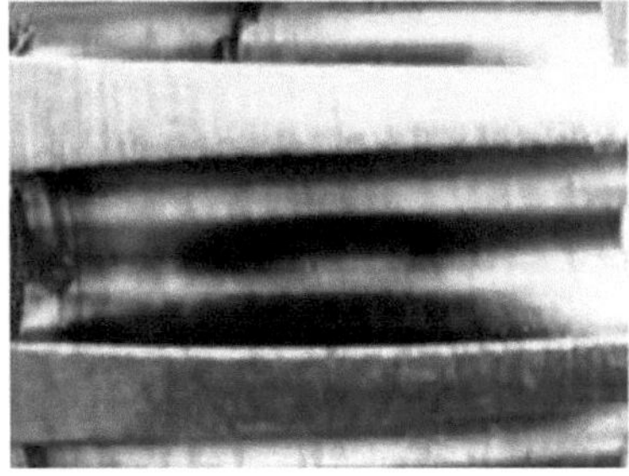

FIGURE 25 - LOCATIONS OF SURFACE TENSION MEASUREMENTS

S The data obtained should be recorded in Excel files in order to assemble the graphs of interest for analysis. It is recommended to evaluate the evolution of the surface tension in each hole measured in the reference positions.

5.2.4. MICROHARDNESS

For the micro-hardness analysis, we relied on the support of a laboratory (LABMAT - Anâlise e Ensaios de Materiais) outside FEMP - UNIMEP, as there is no such equipment in the college's laboratories. These analyses were assisted by technicians from this laboratory and closely monitored by the student responsible for the work.

The equipment used for this analysis was a MICROMET micro-durometer, model 2103, figure 26, which was calibrated according to the MD 036/01 calibrating procedure, which is detailed in section 5.3 of this paper.

FIGURE 26 - MICRO-DUROMETER USED

One of the greatest precautions to be taken to carry out a good microhardness analysis begins with the preparation of the specimen, which must be carried out metallographically. Electrolytic polishing is recommended to avoid hardening of the material on the surface to be analyzed, which would affect the result; if mechanical polishing is used, only a few *microns* of the surface layer should be removed. For small specimens, they should be embedded in bakelite in order to fix them firmly and make their surface perpendicular to the penetrator.

The procedure for carrying out microhardness tests is described in ASTM E384-99:

S Switch on the lighting and equipment;

S Select the desired penetrator, which must be clean and free of static charges;

S Hold the specimen in the device (analyzed sample), keeping its surfaces perpendicular to the working axis of the penetrator, figure 26;

J Set the focus of the microscope using the lowest possible magnification of the image;

J Adjust the intensity of the light and the aperture of the filters to obtain better image resolution, brightness and contrast;

J Select the measurement area to determine the hardness, increase the image magnification and adjust the focus before starting to apply the load;

J Position the penetrator and select the desired load;

J Apply the preload and load for the time indicated in ASTM E 384 - 99;

J After deactivating the charge, switch the device to measurement mode and select the appropriate lenses. Focus the image, adjust the light intensity to obtain maximum resolution and contrast;

J Examine the impression to check that the four corners are in focus and that the diagonals are symmetrical. If these conditions are not as expected, the test should be carried out again on another part of the specimen;

J After obtaining a good impression, measure the diagonals to obtain the average value;

J Calculate the micro-hardness value using equation 1;

J Make a note of the value and repeat as many times as necessary;

It is important to note that there is more modern equipment on the market today, with more advanced systems that allow hardness to be read without the use of formulas. All the operator has to do is position a kind of crosshair at the ends of the two diagonals and the device itself calculates the average of the diagonals and simultaneously displays the value of the hardness measurement analyzed.

When using very low loads (less than 0.3 Kgf), there may be little elastic recovery of the material. They can also produce very small impressions which make it difficult to measure the diagonals (L), due to the difficulty in identifying the corners of the impressions. These factors can cause errors in the test, resulting in lower hardness values than the real ones (TARASOV, APUD SOUZA, 1982).

The machine must be calibrated frequently, mainly because the error in applying the load greatly alters the microhardness value, even with variations of 0.001Kgf for loads less than 0.050 Kgf. The penetrator must have a well-defined tip, otherwise serious errors can occur in the hardness reading (LYSIGHT, APUD SOUZA, 1982).

It is important to remember that usually more than one impression is made per test on the same sample, and to pay attention to the distances between the impressions, which should be on average 3 times the length of the diagonal.

5.3 Characterization of the Equipment Used

This section briefly presents the characteristics of the machines and equipment used in this work:

√ The X-ray diffraction equipment used for the residual stress analysis was the Digital Fastress "Residual Stress Analyzer", model: 1500 SN: 1565-98, manufactured by "Metro Design line USA", with a measurement uncertainty of +/- 5 KPSI, according to the Calibration Certificate issued in 2000. Annex I is attached.

√ For the microhardness measurements, a MICROMET model 2103 microhardness meter was used, which was calibrated according to the MD 036/01 calibrating procedure, with a measurement uncertainty of +/- 1HV. According to Annex II.

√ For image acquisition, a NIKON Microflex HFM, Photomicrographic Attachments metallographic microscope was used, checked and calibrated on February 18, 2000. A Hitachi monochrome CCD camera was attached to a microcomputer with *Global Lab Image View* software.

√ A Starrett Coordinate Measuring Machine, model DCC - RGDC2828- 24, in millimeters, with linear accuracy in X, Y and Z of 0.0051 mm and linear repeatability of 0.0015 mm, according to the manufacturer's catalog, annex III, was used to measure the deviations in dimension and position.

6. RESULTS AND DISCUSSION

This chapter will present and discuss the results obtained from the analyses carried out. These results are based on the procedures described in the previous chapter. The analyses and discussions will be divided into sections, each of which deals with the results individually.

6.1 SHAPE AND DIMENSIONAL DEVIATIONS

The basis for analyzing these deviations was the study of data obtained from measurements made on a CMM, mentioned in Chapter 5 of this work. This data was compiled in electronic spreadsheets ("Microsoft Excel") and through data processing and monitoring of the measured values it was possible to arrive at some results that allowed an evaluation of the behavior of the holes obtained in the specimen over the life of the drill.

With these results, spreadsheets and graphs were generated in order to better display the results obtained, which can be seen in this section of the paper.

The graphs in figures 27 to 29 show the results of the **dimensional deviations (diameter).** Figure 27 shows the graph of the diameter values measured in the holes in the test specimen over the life of the drill bit; figure 28 shows the behavior of the variation in the diameter of these machined holes over the life of the drill bit, with the nominal value of the hole diameter (in this case, 10 mm) as the reference value for the abscissa axis. Figure 29 shows the behavior of the variation in absolute values in relation to the nominal diameter.

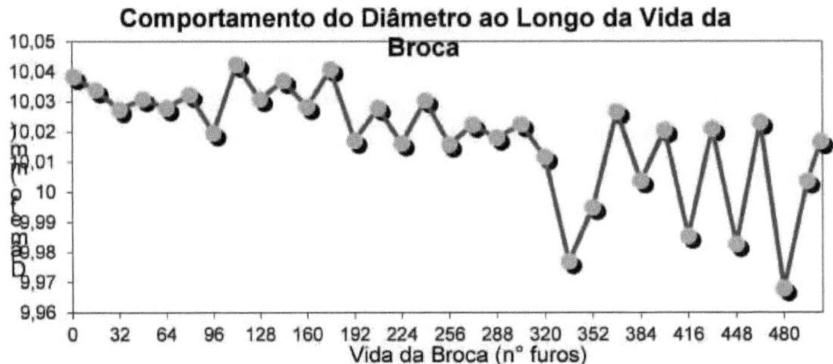

Figura 27 - Diameter values obtained during measurement

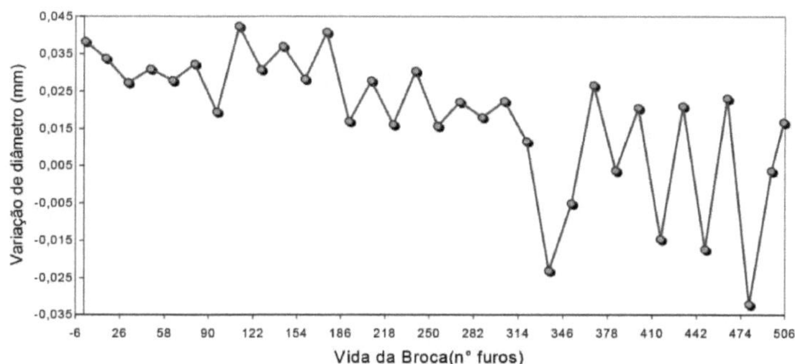

Figura 28 - Behavior of the hole diameter over the life of the drill bit in relation to the nominal value.

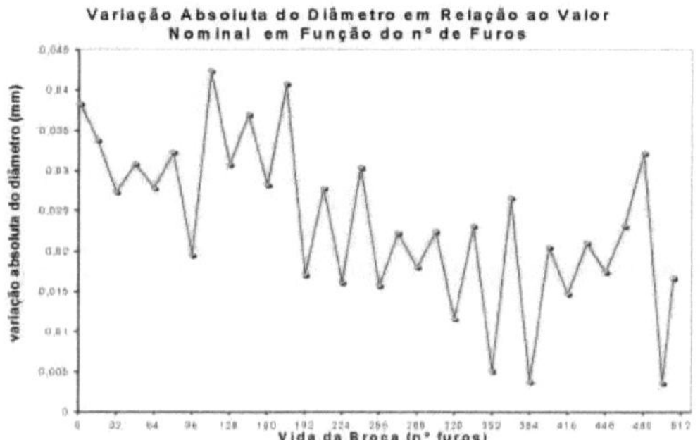

Figura 29 - *ABSOLUTE VARIATION IN HOLE DIAMETER IN RELATION TO THE NOMINAL VALUE MEASURED OVER THE LIFE OF THE DRILL BIT.*

In the graph in figure 27, it can be seen that the diameter values show little dispersion up to approximately hole 300, with all values being above the nominal diameter. After that, the dispersion increases and holes appear with diameters even smaller than the nominal one, and this behavior continues until the end of the drill's life. This behavior may be associated with the evolution of drill wear, which alters the shape of the cutting edges when the hole is drilled.

The graphs in Figures 28 and 29 show that over the life of the drill there is a tendency for the measured diameter to come closer to the nominal value of the holes, which can be credited to the wear of the main cutting edge and the more effective action of the secondary edge, which has a smaller diameter than the main edge. Although this results in more interesting values from the dimensional point of view, this reduction must be analyzed from an economic and operational point of view. With regard to the operational aspect, if the execution of this hole is a stage that precedes another machining operation, such as tapping, this would represent an undesirable oscillation because it directly affects the stress conditions suffered by the tools during the process, which could lead to their breakage and the loss of the part or an increase in the time it takes to recover it.

Excessive wear of the main edge is also not interesting from the point of view of reusing the drill, as it compromises the coating layers and prevents regrinding. The composition of the carbide substrate is also altered, as it is possible that some of the elements may diffuse as a result of direct contact with the workpiece. A chemical analysis carried out on one of the drills used in the tests showed that the proportion of the substrate's element content had been altered, as shown in Table 2.

TABLE 2 - PERCENTAGE DISTRIBUTION OF THE ELEMENTS FOUND IN THE Chemical ANALYSIS OF the New Bark (%INITIAL) AND AT THE END OF LIFE (°%OFINAL)

Substrate substrate	Initial %	% final
Ti	0,04469	0,05434
Co	0,17538	0,47715
W	0,77993	0,40770
Fe		0,06081

The most significant changes are a reduction in the percentage of W and an increase in the percentage of Co, as well as the appearance of residual Fe, probably caused by impregnation of the machined material (Annex IV). As a result, the drill cannot be reused at the end of its life and must be discarded. As it is a relatively expensive tool, due to its constructive characteristics and composition, its disposal is economically undesirable.

With regard to the **perpendicularity** of the holes, figure 30 shows the behavior observed over the life of the drill. It can be seen that at the beginning of the drill's life, the values obtained have little dispersion of variation; only after the 100th hole does the dispersion begin to increase, rising significantly after the 300th hole. In absolute values, as shown in the graph in figure 31, this increase in the dispersion of deviations is more evident. It is interesting to note that this behavior occurs at almost the same time as the reduction in diameter.

Figura 30 - Variation in the perpendicularity of the holes over the life of the drill bit.

Figura 31 - Absolute variation in the perpendicularity of the holes over the life of the drill bit.

In the same way as in the case of diameter, these behaviors of increased dispersion of the variation in perpendicularity can be associated with excessive wear of the main cutting edges of the drill bit; it is also likely that the deviation occurs due to asymmetrical wear of the main edges of the drill bit, as shown in figure 32.

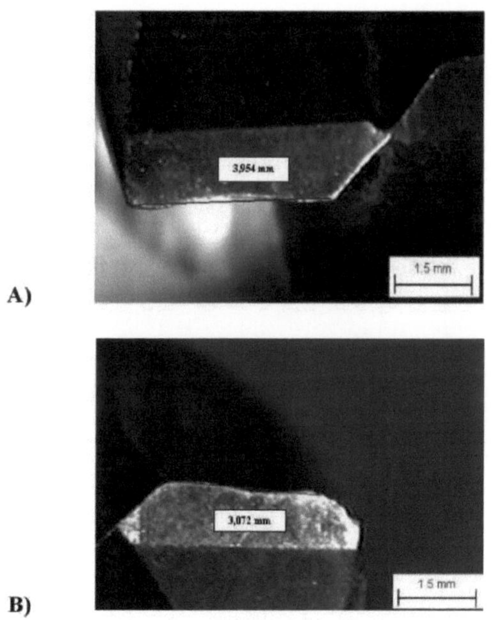

FIGURE 32 - FIGURES OF THE MAIN CUTTING EDGES MEASURED

This uneven wear, identified in figure 32, causes different stresses on each side of the drill bit and, consequently, different moments on each side, resulting in a misaligned twisting moment in relation to the center of the drill bit and inducing a deviation in the direction of the hole.

The imbalance in cutting forces could also be one of the reasons why the drills broke in some tests. Wear on one of the main edges was relatively greater than on the other, causing the drill body to twist.

With regard to the analysis of the **deviations in the location of the holes** along the surface of the specimen, data was collected on the individual positioning of the holes in both the longitudinal direction (X coordinate) and the transverse direction (Y coordinate). This data is translated into incremental mode and is shown in the graph in figure 33.

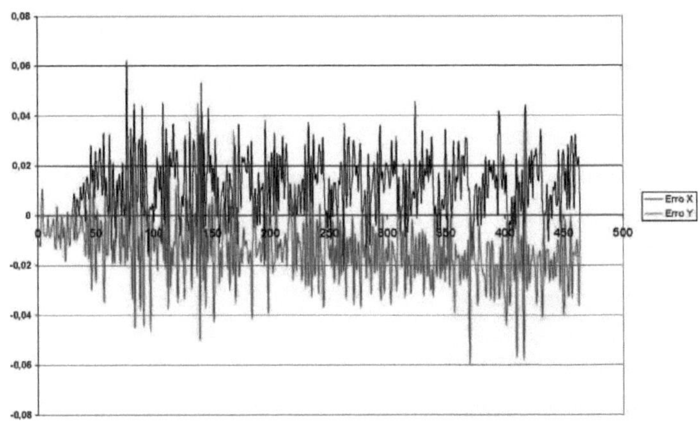

Figura 33 - Variation of hole location on the X and Y axes

Given the large number of samples represented in the graph in figure 33, it is difficult to interpret the data. Nonetheless, deviation trends can be seen on both the X and Y axes as the drilling process moves towards the end of the drill bit's life.

To make it easier to visualize and interpret the data, the measured and nominal positions of the holes were mapped. Figure 34 shows this mapping, which makes the deviations from the X and Y axes found in the holes more obvious.

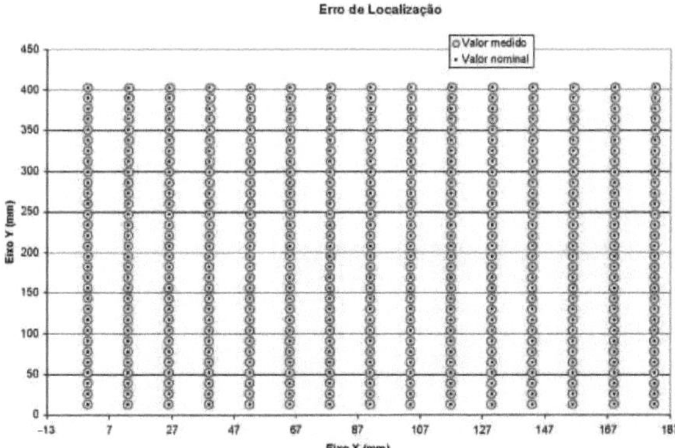

Figura 34 - VARIATION IN DISPLACEMENT OF THE X AND Y AXES

It is important to note that for a more in-depth analysis of these location errors, both the positioning repeatability error of the machine tool that performed the drilling and the positioning error of the CMM that performed the measurement must be taken into account. Even so, it is possible to state that throughout life there is a tendency for this type of deviation to increase, which can be credited to the asymmetrical wear of the cutting edges.

6.2 MICROHARDNESS

The microhardness analysis was carried out using the Vickers microhardness method, following the instructions in the ASTM E384-99 standard. Some of the steps in the standard procedure did not need to be followed, given the modern characteristics of the equipment used for reading microhardness values.

To carry out this analysis, a mixture of normal microhardness measurement and gradient analysis was applied, a practice commonly used by technical staff in laboratories specializing in materials analysis.

The analysis was therefore carried out on a sample of 16 holes, following the logic of gradient analysis: microhardness measurements were taken in different positions and in regions close to the surface of the samples analyzed. A heat gradient analysis was not actually carried out, as the distances adopted between one impression and the next were not controlled as required by the standard governing this type of analysis, not least because it would be a little outside the main focus of the work.

The load used to measure the samples analyzed was 0.025 kg, chosen according to the procedures of the ASTM E384-99 standard. According to this standard, the distance between impressions should be approximately three times the average diagonal dimension; for this reason, it was necessary to adopt an inclination in the measuring line in order to concentrate the maximum number of measurements at small distances from the sample surface.

To better understand this process, two images were taken of samples that had undergone microhardness analysis at different stages of drill wear (hole 416 and 160), which can be seen in figure 35.

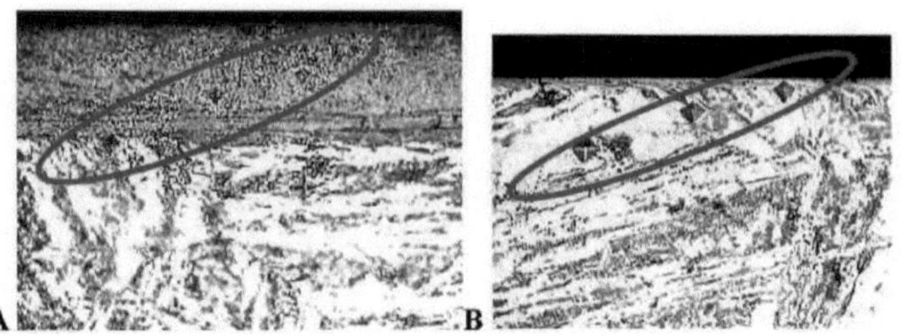

Figura 35 - **A)** SAMPLE 416B**)** 160

The images in figure 35 show how the prints were arranged. It is also possible to see the difference in the size of the impressions; the closer to the surface, the smaller the size of the impression, and consequently, the smaller the diagonal size and also the microhardness value found, given that for a constant load the size of the impression obtained is directly proportional to the magnitude of the microhardness found.

The results of this analysis were used to generate graphs and Excel spreadsheets to help study the behavior of the microhardnesses found in the specimens throughout the test. Part of these graphs are included in this section of the paper in order to better illustrate the phenomena identified in the structures of the samples analyzed. It is important to note that in order to begin the analysis, some measurements were made on the samples in regions far from the surface and free of apparent structural changes, in order to obtain the average microhardness of the material analyzed, with which the reference value (310 HV) was defined for comparative purposes and to prove changes in behavior.

The graph in figure 36 was generated with the values of the highest hardnesses found in the samples. The impressions that gave rise to these micro-hardness measurements were generally concentrated in a range within the sample that did not exceed 0.02mm from the surface of the samples analyzed.

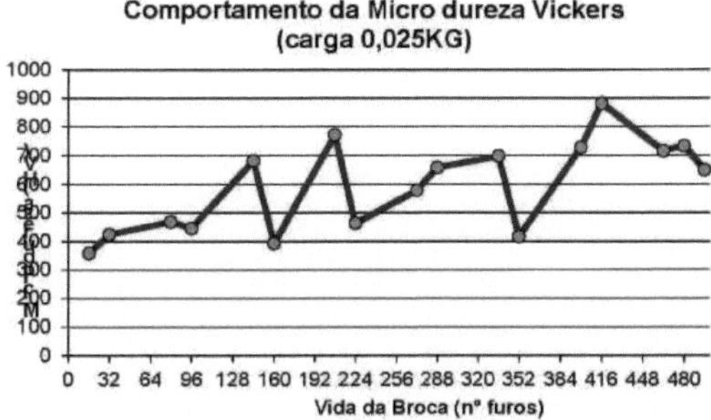

Figura 36 - Microhardness behavior in relation to the end of life of the drill.

It can be seen that there is a tendency for the microhardness to increase over the life of the drill bit, reinforcing the theory that the walls of the hole suffer some kind of thermal and/or mechanical influence due to the wear of the main cutting edges.

This thesis is confirmed by analyzing the graphs in Figures 37 and 38, which show the behavior of the microhardness along the surface and subsurface of the samples. The variation in microhardness in the regions analyzed indicates that there has been a change in the mechanical characteristics of the material in the surface and subsurface layers of the samples, due to stresses during the drilling process.

This analysis was carried out on all the samples subjected to microhardness measurements, which amounted to a total of 16 analyses, with an average of 5 measurements being taken for each one. Figures 37 and 38 are part of this set of analyses and refer to samples from holes number 16 and 480 respectively, as indicated in the figures. The rest of the graphs for this analysis can be found in Annex V.

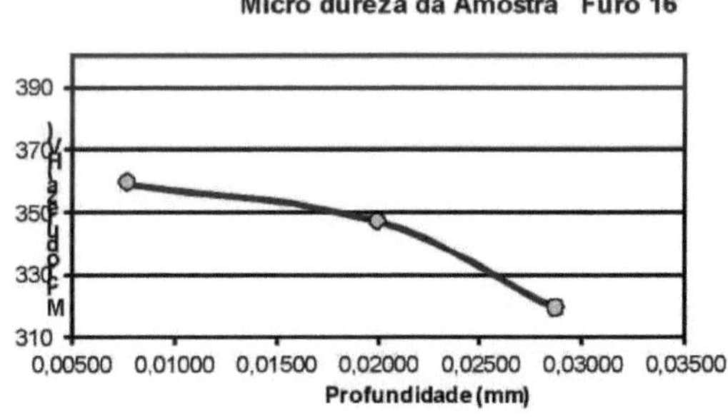

Figura 37 - Microhardness analysis of sample 16

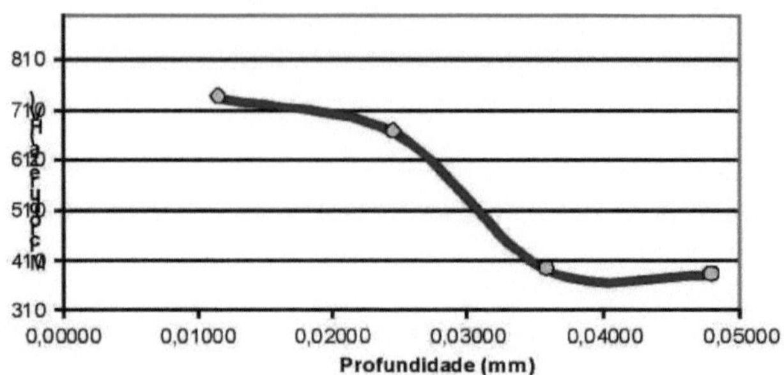

Figura 38 - *SAMPLE MICROHARDNESS ANALYSIS 480*

As in figure 37, the graph in figure 38 also provides data that consolidates the hypothesis that a layer has appeared, possibly as a result of a stress generated during the process of drilling the hole. What differs in these samples is the proportion of variation in the microhardness behavior and the propagation of the layer towards the core of the part. These characteristics are more pronounced in the graph in figure 38, which corresponds to the analysis of the surface and subsurface of the sample referring to hole number 480, a situation of drill bit wear close to the end of its life, unlike the situation found in the analysis in figure 37, a situation at the beginning of the drill bit's life.

Along the same lines of proposing that the existence of the layer affected by the drilling process is a good indication, the occurrence of changes in mechanical properties is shown in figure 39: the graph shows the general behaviour of the samples analyzed with regard to the depth at which the measured microhardness values stabilize, the region where the reference value is reached (310 HV).

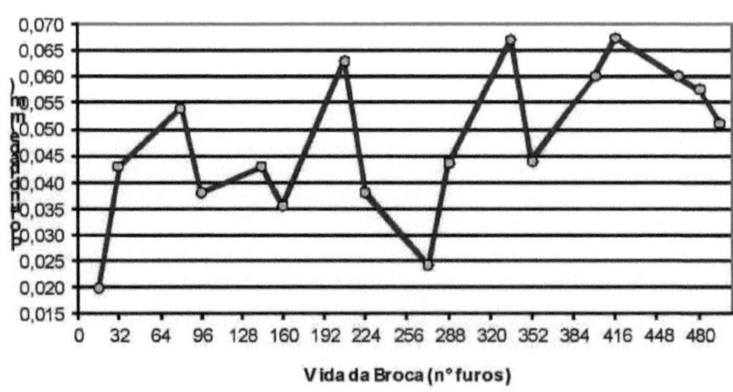

Figura 39 - *MICROHARDNESS STABILIZATION POSITIONS*

It is possible to make a projection of the increase in the depth of the affected layer from the location of the stabilization of the microhardness behavior, which becomes more distant from the surface of the sample as the process moves towards the end of the drill's life.

The analysis of the microhardnesses carried out and represented in this item provides a strong indication that there is indeed a heat-affected layer on the surface and subsurface of the samples analyzed. This layer could contribute to compromising the surface integrity of the samples.

6.3 Heat Affected Layer (HAC) Formation Analysis

With regard to the integrity of the surface and subsurface layer of the samples, assessments of the appearance of CAC were made using image analysis, as described in Chapter 5.

As a result of this image analysis, graphs are presented showing the behavior of the dimensions of the BACs that occurred in the samples throughout the test. These graphs help to interpret the behavior of BAC formation as a function of drill life. Along with these graphs, some images of the samples analyzed are presented in order to demonstrate not only numerically, but also visually, the phenomena occurring in the material structure of the samples subjected to the drilling test.

Just as in the analysis carried out in item 6.2, the analysis of this item also detected the formation of an affected layer which repeats the same behavior identified in the analyses of that item.

As a starting point for this analysis, a sample of the specimen material was prepared, taken from a region far away from the regions involved in the test. From this sample, its microstructure was acquired in image form in order to serve as a visual comparison parameter between unaffected and affected microstructure, as shown in figures 40 and 41, respectively.

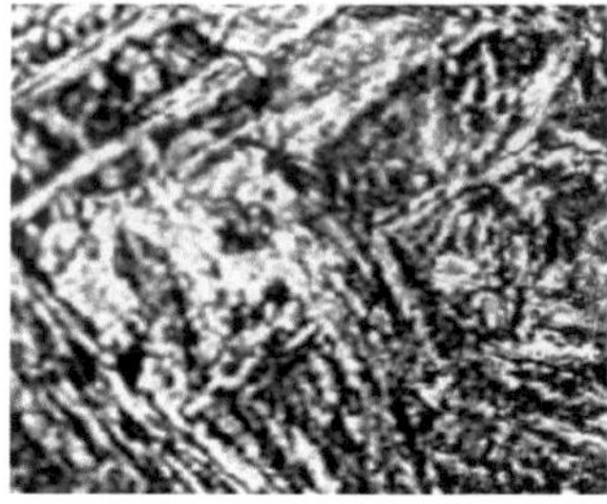

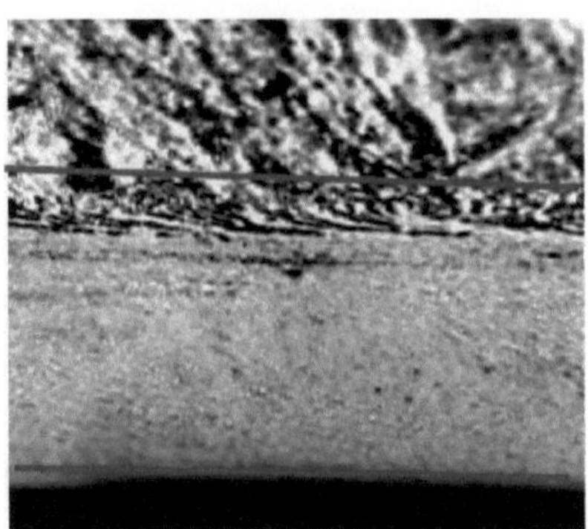

Figura 41 - *AFFECTED MICROSTRUCTURE SAMPLE*

Figure 41 shows an image of CAC, which has a denser microstructure and lighter shades than the microstructure shown in Figure 41 and the rest of the microstructure shown in Figure *42,* above the region delimited by the lines (SHAW, 1994; VIEIRA, 1999).

The main purpose of presenting Figures 40 and 41 was to emphasize the differences between the samples with the affected microstructure and the unaffected sample, in order to give the reader a better understanding of the results and data presented in this section.

This first contact with the image of an affected microstructure proved the real existence of CAC, which consolidated and justified the need to carry out an image analysis; the values obtained from this analysis are presented below, starting with Table 3, which shows the CAC depth values obtained from the analyses. The samples were selected according to the measurement strategy adopted previously (16 in 16 holes).

The samples were selected according to the strategy used to obtain the holes (16 out of 16 holes). With regard to the values shown in table 3, indications can be seen of behavior similar to that observed in the analysis of the previous item, where there is a projection of the values as a function of the wear of the drill bit (end of life).

TABLE 3 - TABLE OF CCS MEASUREMENT VALUES

Hole (n°)	CA C (mm)	Hole (n°)	C AC (mm)
16	0,01	272	0,034
32	0,015	288	0,031
48	0,014	304	0,044
64	0,021	320	0,029

80	0,026	336	0,064
96	0,032	352	0,042
112	0,023	368	0,038
128	0,042	384	0,039
144	0,031	400	0,067
160	0,036	416	0,038
176	0,047	432	0,046
192	0,041	448	0,038
208	0,057	464	0,068
224	0,022	480	0,049
240	0,02 3	496	0,051

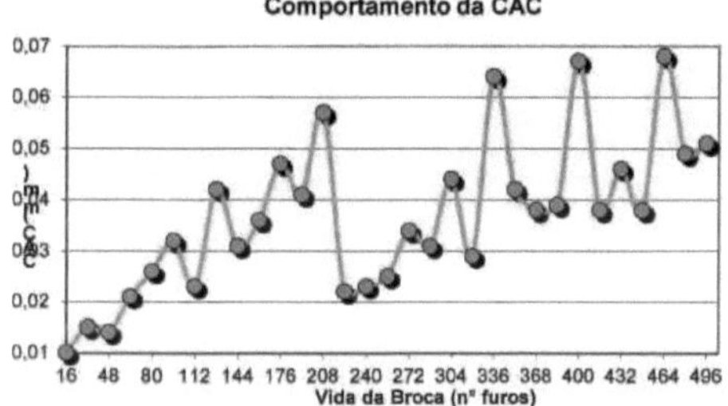

Figura 42 - *BAC BEHAVIOR THROUGHOUT THE TEST*

The graph in figure 42 shows that there was a significant drop in the CAC value around hole 208. This drop may be the result of a pause in the drilling test which allowed the drill bit and the specimen to cool down. When the test was resumed, the CAC began to form again and showed an upward trend.

The CACs found and analyzed have depths ranging from 0.01 mm at the start of the test (hole 16) to 0.051 mm at the end of the test, the last hole analyzed in the specimen (hole 496). Some values are even greater than those identified in the last hole, with the maximum depth measurement analyzed being the CAC for the sample in hole 464, which reached a value of 0.068 mm.

During this analysis, images were taken of all the samples, some of which are shown below. These images show the value of the depth of the CAC, measured in mm; recorded on the images themselves, as a facilitator for the reader to interpret the results. These images can be seen in figures 43 to 48.

The other images are shown in Annex IV. It is worth noting that the CAC measurement assumed the region in which the grains of the material's microstructure undergo deformation and are oriented in the direction in which the cut occurred. This deformation is clearly visible in Figures 45, 46 and 47.

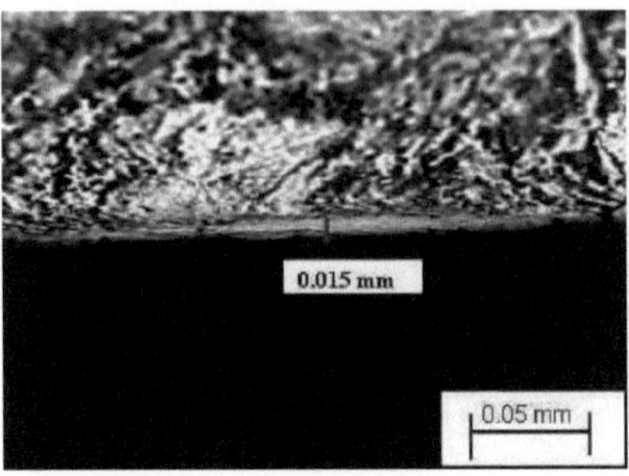

FIGURE 43 - CAC OF HOLE 16

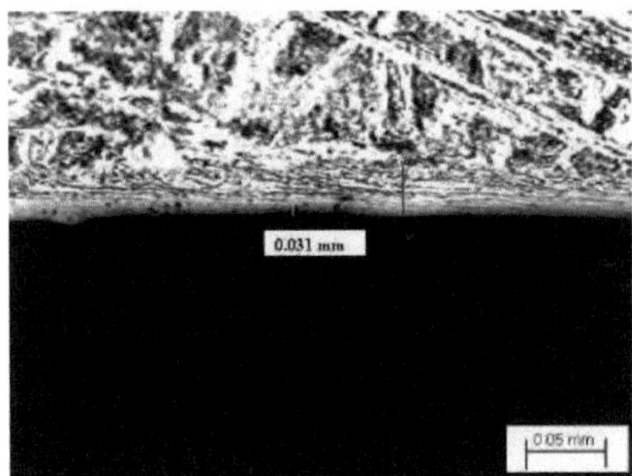

FIGURE 44 - CAC OF HOLE 144

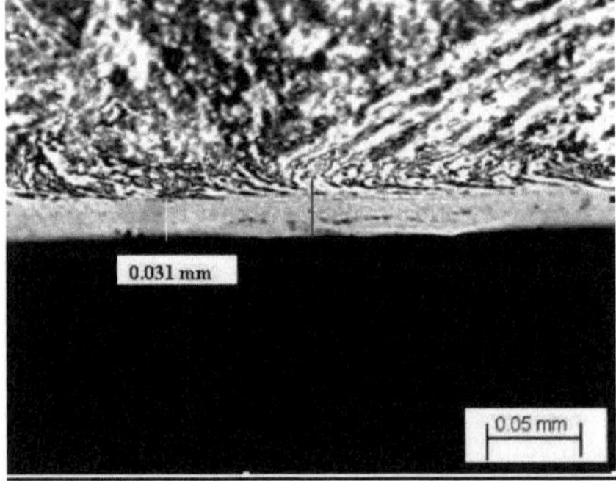

FIGURE 45 - CAC OF HOLE 272

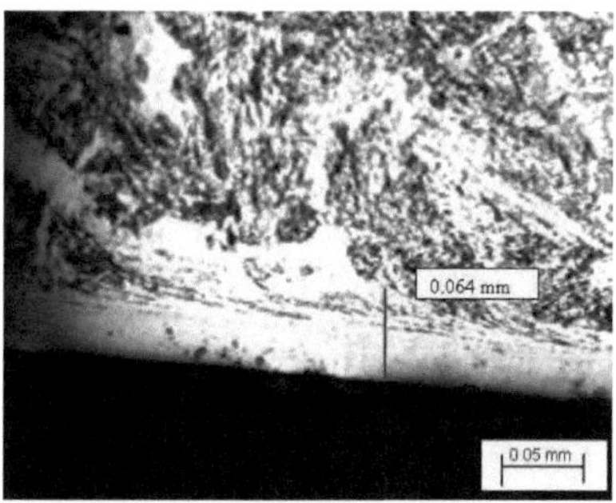

FIGURE 46- CAC OF HOLE 336

FIGURE 47 - CAC OF HOLE 464

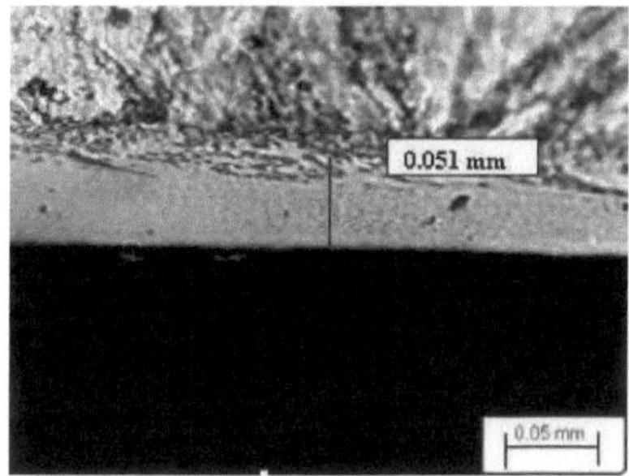

FIGURE 48 - CAC OF HOLE 496

By monitoring the images of the CACs illustrated in figures 43 to 48, we can observe the behavior of the microstructure of the surfaces and subsurfaces of the samples analyzed. This behavior can be described as an evolution in the values identified for the depth of the CAC in the samples near the end of the drill bit's life (hole 464, depth of the CAC = 0.068mm, figure 47), in relation to the samples near the beginning of the drill bit's life (hole 16, depth of the CAC = 0.015mm, figure 43, and reaching the end of the drill bit's life (hole 496), with a depth of CAC = 0.051mm, showing a small decline in the behavior of CAC evolution, as shown in figure 48.

Also with regard to this set of samples, a comparison was made between the survey of the depth of the CAC throughout the test using image analysis resources and the data obtained from the microhardness tests. To better understand this relationship, the graph in figure 49 shows the two data sequences simultaneously, measuring the depths of the layers identified in the two analysis methods.

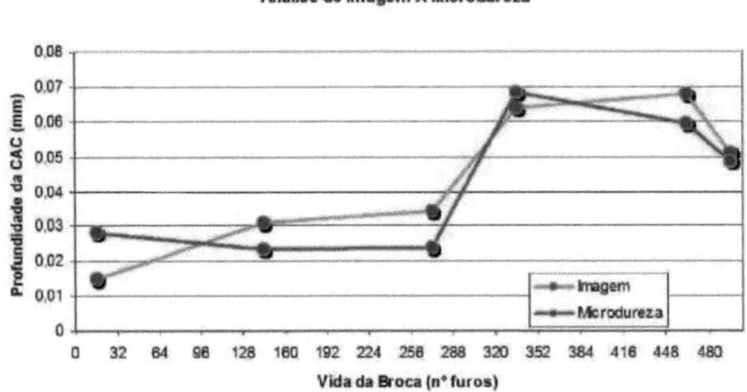

Figura 49 - *IMAGE ANALYSIS X MICROHARDNESS ANALYSIS*

The graph in figure 49 shows a similar trend of behavior in both analyses (microhardness and image analysis). One result ends up supporting the other, confirming that some kind of transformation has indeed taken place in the microstructure of the specimen material after the drilling test.

To complete the analysis in this section, we decided to repeat the strategy adopted above, in which we compared the data from the microhardness analysis carried out in the previous section with that obtained from the image analysis. However, this time a comparative analysis was carried out with all the samples analyzed in the previous section (17 samples), confirming the existence of a correlation between the two behaviors analyzed (CAC and Microhardness).

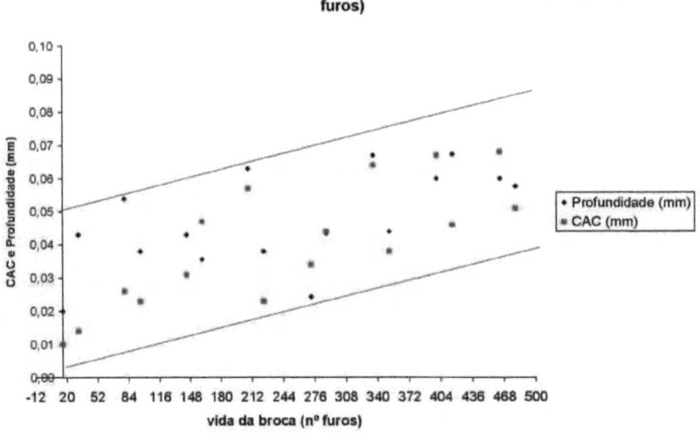

Figura 50 - IMAGE ANALYSIS XANALYSIS OF THE MICROHARDNESS OF THE ASSEMBLY

To conclude these comparative analyses, figure 56 gives the reader an idea of how the mechanical microhardness characteristic behaves along the surface and subsurface of the samples analyzed here. To this end, an image of the CCS analysis has been superimposed on a graph with the microhardness analysis data for the same sample (sample from hole 464).

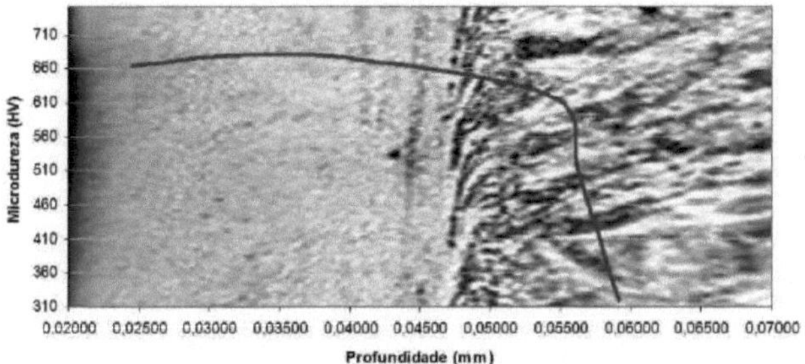

Figure 51 - Composition of analyses

It is clear to see that in the clearest region of the image, where the CAC has formed, the microhardness has a higher value. In the region where the image shows the microstructure without deformation (from a depth of 0.05 mm), the microhardness decreases significantly until it reaches the reference value (310 HV). And in the intermediate region (between 0.045 and 0.05 mm), the microhardness decreases slightly, but still remains at high values due to the deformation of the grains in the structure of the material (a probable hardening of the material).

6.4. STRESS ANALYSIS

The results obtained show the behavior of the surface stresses both along the reference surface of the specimen, which had no direct contact with the tool but suffered stresses from the test, and on the machined surface (hole wall). The positions at which the measurements were taken are shown in figure 25 and are identified as A (reference surface of the piece), B and C (wall of the hole), according to the data in chapter 5 of this work.

The graph in figure 52 shows the results and the behavior of the surface tension in the three positions indicated throughout the life of the drill bit, as well as the reference value of the tension, measured in a region of the workpiece that was not subjected to the drilling process.

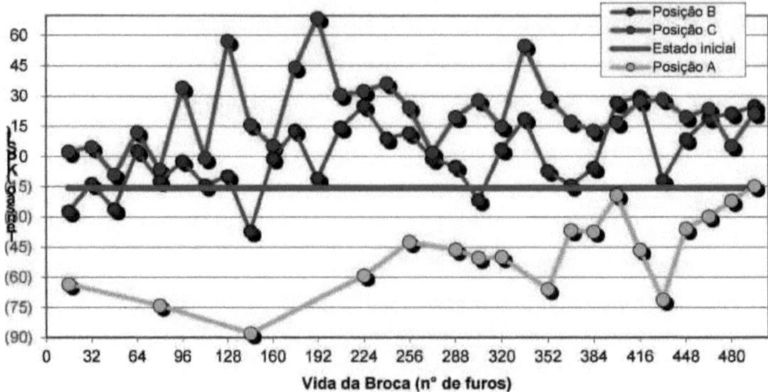

Figura 52 - *CORRELATION BETWEEN VOLTAGE DATA*

Analysis of this graph (Figure 52) shows that:

- On the surface of the specimen (position A) the stresses were all compressive, but with a tendency to reduce towards the neutrality line (zero stress);

- In position B, the stresses were distributed between compression, mainly at the start of the drill bit's life, and tension, at the stage when the drill bit was halfway through its life. The graph in figure 53 shows the proportion in which this distribution occurred;

- In position C there was a greater concentration of tensile stresses. The graph in figure 54 shows the proportion in which this distribution occurred.

Figura 53 - Distribution of voltage type in position B

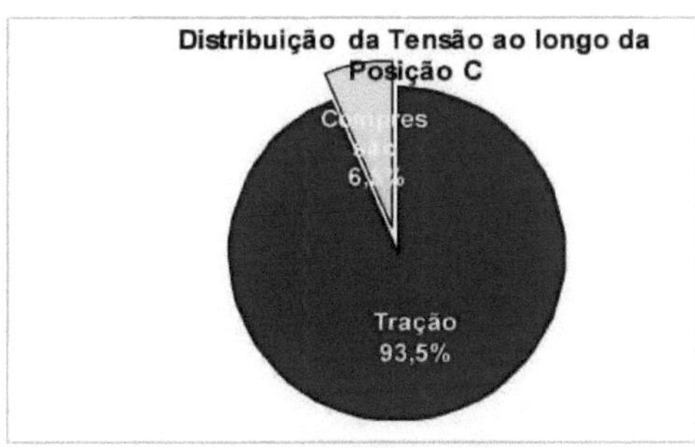

Figura 54 - *DISTRIBUTION OF VOLTAGE TYPES*

In all cases, it can be seen that there is a tendency for the stress to grow towards the tensile state, indicating that the loss of cutting of the main edge of the drill induces the workpiece into a state of undesirable tensile stress, as the hole could be a point at which a crack or other type of flaw could appear.

This behavior of the stress variation along the specimen can also be taken as an indication that something is probably happening to the integrity of the material, since practically all the values are very far from the reference value of the initial stress state of the specimen.

The differences in stress state behavior can be linked to differences in heat transfer rates, given the fact that the positions analyzed are distributed in different regions of the specimen, as shown in figure 25 in section 5.2.3 of this paper. In these different regions of the specimen there are notable differences in terms of material cooling behavior, given the variables involved (material thickness, exposure to air, among others).

6.5. RELATIONSHIPS BETWEEN ANALYSIS RESULTS

The purpose of this section is to carry out an exploratory analysis of the data obtained from the different tests carried out on the specimen.

Firstly, a comparative analysis is made of the behavior of the diameter of the holes over the life of the drill with the behavior of the CCS, verified by both microhardness and image analysis (figure 55).

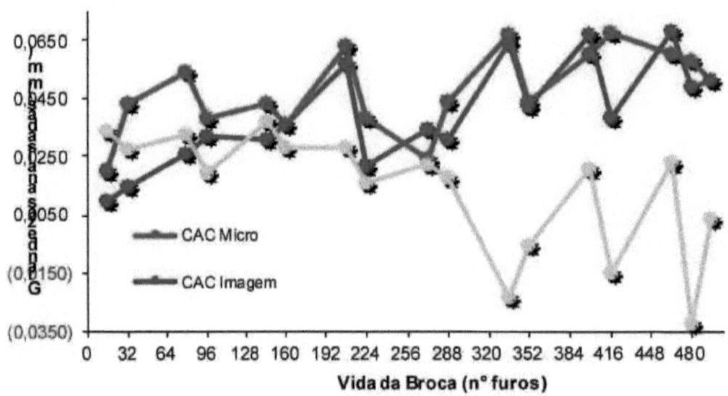

Figura 55 - COMPARISON OF CCS BEHAVIOR AND DIAMETER

The graph in figure 55 shows a behavior with little dispersion of the related quantities up to a certain point in the process, around hole 300; from then on, the quantities analyzed undergo a significant change in their respective behaviors, and the reduction trend already identified earlier in the diameter is associated with an increase in the CAC. This variation in behavior may be a strong indication of the existence of an end-of-process moment that is different from the end of the drill bit's life.

Figure 56 shows the behavior of the measured microhardness and surface tension at points B and C of the samples. The microhardness is expressed here in HRC (Rockwell C) to make it easier to see the graph. It can be seen that the behavior of the tension in position B, also around hole 300, becomes very close to the behavior of the microhardness. In position C, however, this indication of the closeness of the behavior to the microhardness is less clear. Even so, these three quantities could also be indicating the existence of a process end point.

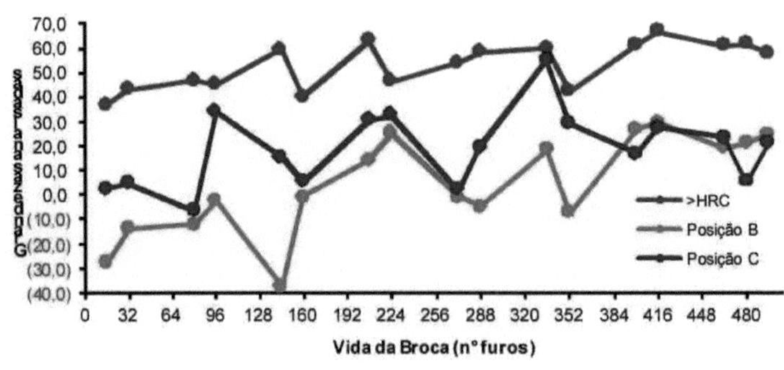

Figura 56 - *COMPARISON OF MICROHARDNESS AND STRESS BEHAVIOR*

In order to more clearly identify the existence of a process end point through the possible correlations between surface tension and microhardness, an analysis was also carried out of the behavior of the tension in comparison to that of the formation of the CAC observed by both image analysis and microhardness (figure 57).

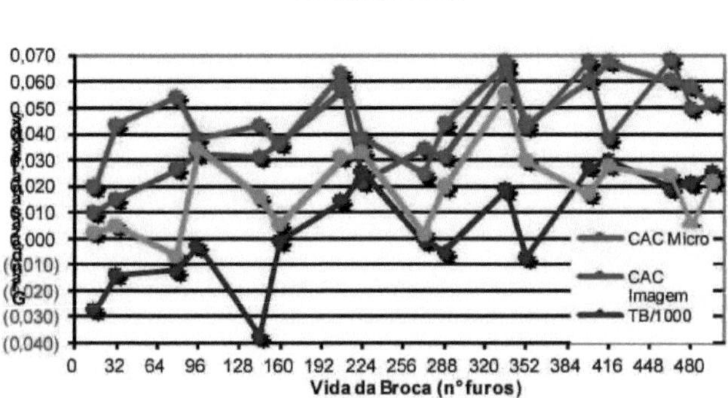

Figura 57 - Comparison of the behavior of tensions and CACs

In the graph in figure 57, it can be seen that at the points previously indicated as the probable end of the process - around hole 300 - the stresses tend to rise and the CAC reaches high values that could compromise the part when it is used.

Finally, by analyzing the behavior of the variation in diameter together with the variation in perpendicularity, the existence of this end-of-process moment is clearly identified when both quantities begin to show a very large dispersion in their respective behaviors (figure 58).

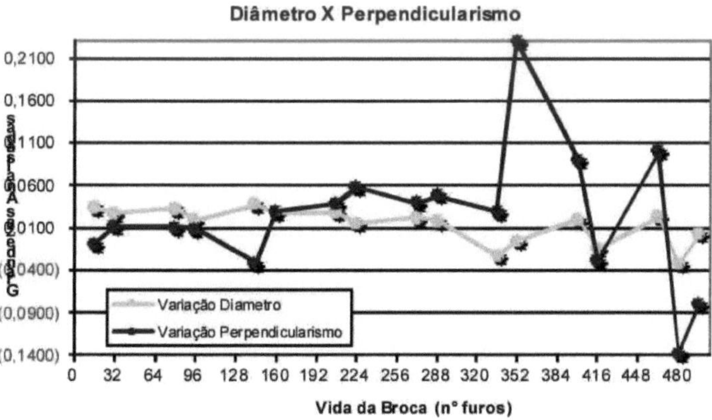

Figura 58 - COMPARISON OF THE BEHAVIOR OF DIAMETER AND PERPENDICULARITY

Despite the trend towards a reduction in diameter, it is noticeable that this is accompanied by a large variation in perpendicularity, which is undesirable in functional elements that must serve as a positioning reference or as a passage for fasteners.

7. CONCLUSIONS AND SUGGESTIONS FOR FUTURE WORK

Based on these results, we conclude that:

√ There is a limit to the use of the drill other than the end of life. This limit can be considered as the end of the process, and can be indicated by monitoring the MmdQ;

√ An analysis of the diameter alone can lead to erroneous conclusions, as it acts as an approximation of the nominal value and does not indicate the existence of other deviations from the MmdQ that may occur;

√ The variation in perpendicularissimo, as well as the variation in the deviation of the location of the holes are good criteria for indicating the end of the process, as they do not require destructive analysis;

√ Drill wear influences deviations in perpendicularity and location, compromising the final results of the process;

√ Even before the end of the process, the formation of the CAC was identified by both the microhardness method and the image analysis method. The appearance of this layer can affect the use of the part if another normalizing operation is not carried out on the part;

√ The use of optimized process parameters induces the formation of CAC and compromises the surface integrity of the holes obtained. It is again clear that using the drill until the end of its life is not advisable;

√ The use of optimized drilling and without cutting fluid is feasible up to the end-of-process point. Using the drill beyond this point (or up to the end of its life) can induce problems in the MmdQ that can result in failure in the use of the part (fracture or crack) or in another manufacturing phase being carried out to correct the problems identified: deviations in perpendicularity, position, the appearance of CAC and the existence of residual stresses, especially tensile stresses;

√ By identifying the end-of-process point, wear conditions are achieved that make it possible to reuse the tool, which is very interesting from a financial point of view, as it makes it possible to use the optimized parameters (high productivity) and reuse the drill bit (saving on tool purchases).

8. BIBLIOGRAPHICAL REFERENCES

AGOSTINHO, O. L. RODRIGUES, A. C. S. LIRANI, J. Tolerances, Adjustments, Deviations and Dimension Analysis. Sao Paulo, 1977. Edgard Blucher of the University of Sao Paulo. 296p.

ASTM, E 384-99, Standard Test Method For Microindentation Hardness Of Materials. P.409 - 424.

BALAKSHIN, B. Fundamentals of Manufacturing Engineering. Chap. I - IV, translated by ROZENFELD, H. Apostila Laboratório de Ferramentas, Sao Carlos 1983. 194p.

BRAGA, D. U. The Technique of Minimum Quantity of Cutting Fluid Applied in the Drilling Process of an Aluminum Silicon Alloy. Faculty of Mechanical Engineering, State University of Campinas, 2001, 177p. Thesis (Doctorate).

BRAGA, D. U. DINIZ, A. E. COPPINI, N. L. Utilization of the Minimal Mist Lubrication Technique in the Drilling of Silicon Aluminum Alloy SAE - 323. Usinagem Brasilwww.usinagem-brasil.com.br/artigostecnicos . Accessed November 2001.

CARDOSO, J. C. M. Estudo de Caso para a Implantaçao De "Manufatura Classe Mundial" e Proposta de Conceito Para "Empresa Classe Mundial. Faculty of Mechanical and Production Engineering, Methodist University of Piracicaba, 2000, 113p. Dissertation (Master's Degree).

CRUZ, A. R. Jr. Processo de Furaçao Escola de Engenharia de Sao Carlos Departamento de Engenharia Mecânica. Sao Carlos, 1978 (handout).

DAMASCENO, D. Residual Stress Analysis after Turning and Grinding of Hardened ABNT 52100 Steel. Faculty of Mechanical Engineering, State University of Campinas, December 1993, 97p, Dissertation (Master's Degree).

DEGARMO, P. E. Materials and Processes in Manufacturing. Ed. Macmillan Publishing. Printed in the United States Of America, 1979. Fifth edition, 857p

DIETER, G. E. Mechanical Metallurgy, Ed: MacGraw-Hill, Inc., New York - USA, 1961-1976. 248p.

DINIZ, A. E, COPPINI, N. L, MRCONDES, F. C., Tecnologia de Usinagem dos Materiais. Ed. Art Liber Editora, SP, 1999 1° ed.244 p.

FARAGO, F. T. CURTIS, M. A. Handbook of Dimensional Measurement, Ed. Industrial Press Inc. Third Edition. 1994, p.580.

FERRARESI, D. Fundamentals of Metal Machining. Sao Paulo: Edgard Blucher Ltda., 1977, 751p.

FREIRE, J. M. Tecnologia Mecânica - Sawing and Drilling Machines, volume 2. Rio de Janeiro, 1978. Ed: Livros Técnicos e Cientificos.230p.

GRIFFITH, G. K. Measuring & Gaging Geometric Tolerances. Ed. Paramount Communications Company Englewood Cliffs, New Jersey, 1994 p 1 - 301.

GUNTER SPUR, H. C. HOK TIO, T. Damage to the Surface Layer in the Machining of Advanced Ceramics, Machines and Metals, January, 1991. Ed Aranda. p 64-72.

HEISEL, U. & LUTZ, M., Coolant and lubricant research, Machines and Metals, May 1998. Ed. Aranda. p. 40-49

JACOBUS, K. DEVOR, R. E. KAPOOR, S. G. Machining - Induced Residual Stress: Experimentation And Modeling. Transactions Of the ASME - American Society of Mechanical Engineers, Vol. 122, February 2000. p20 - 31.

KALHOFER, E. Dry Machining - Principles and Applications. Proceedings of the 2nd High Technology Seminar "Machining with High Cutting Speed and High Precision". Methodist University of Piracicaba, July 1997, Santa Bàrbara D'Oeste, Sao Paulo, 5p.

KAMMERMEIER, D., BORSCHERT, B. KAUPER, H. SCHENEIDER, M. Drilling Without Cooling: Only Ecological Reasons? . Metal Mecânica, Sao Paulo, year XViii, April/May 2000, p6269. and Internet: WWW.usinagem-brasil.com.br/artigostecnicos, Accessed November 5, 2001.

KLOcKE, F. & EiSENBLATTER, G., Presented at the Opening Session Dry cutting, Annals of the ciRP, v. 46 (2), pp. 519-526, 1997.

KONiG, W. RUMMENHOLLER, S. Industries Are Having to Orient Their Production Processes Ecologically. Machines and Metals, Vol. 387, p22-29, April 1998. Ed Aranda.

LiMA, A. ViEiRA, M. JR; Brief Study of Macro Quality Determinats of Pieces Obtained From Optimized Drilling Process. Anais do Vi Encontro de Mestrando em Engenharia May 17 - 19, 2002. Sao Pedro. SP.

LiMA, A. ViEiRA, M. JR; LiBARDi, R; cANciLiERi, H. A. Study of the Macro and Micro Determinants of the Quality of Sintered Parts Subjected to the Grinding Process. Proceedings of the XXi Encontro Nacional de Engenharia de Produçao, ENEGEP 2001, October 17 - 19, 2001, Salvador. BA

LiRANi, J. introduçao a Metrologia industrial, Apostila, Sao carlos, 1985, Publication 053/87. 1st issue p. 78.

LYSiNGHT, V. E., Metal Progr., (78), p93, 1960.

MARiN, J. M., Mechanical Behavior of Engineering Materials. Ed: Prentice - Hall, inc., USA, 1960. 318p.

MATEOS, A. G. Tolerâncias e Ajustes, transl. Neiva, A. C. Ed. Poligono. Sao Paulo, 1974. p161 - 405.

MiRANDA, G. W. A.; cOPPiNi, N. L.; BRAGA, D. U.; DiNiZ, A. E., Contribution to the Drilling Process with Coated Carbide Drills, Brazilian Congress of Manufacturing Engineering - COBEF 01, April 2001, Curitiba - Paranà. CD of the Proceedings of COBEF 01

MULLER, P. Drilling and tapping tools with HSC and without coolant. In: O Mundo da Usinagem, Sao Paulo/SP, Ed. Sandvik, January 2000. p. 13 to 17.

NAKAGAWA, H. Minimal Lubrication Does Not Harm the Environment, Magazine: Maquinas e Metais, ED. Aranda, August 2000, p 40-49.

NOVASKI, O. J. RIOS, M. Vantagens Do Uso De Fluidos Sintéticos Na Usinagem. Internet: WWW.usinagem-brasil.com.br/artigostecnicos, Accessed November 2001.

NOVASKI, O. J. Introduction to Mechanical Manufacturing Engineering. Edgard Blucher Ltda. 1998, p. 47 - 80.

PALLEROSI, C. A ET AL., Durability of Cutting Tools Under True Conditions, Proceedings of CANCAM 91 -Canadian Congress in Applied Mechanics, p. 173-175, Canada, 1991.

PUNCOCHAR, D. E. Interpretation Of Geometric Dimensioning And Tolerancing. Second edition. Ed. Library of Congress Cataloging in Publication Data, Industrial Press Inc. New Yourk, printed in the United States Of America, 1996. P 1 - 97

RODRIGUES, M. A. P. ABRÂO, A. M. Castor Oil Derivatives as Cutting Fluid. Machines and Metals, Vol. 400, p104 - 112. May 1999. Ed. Aranda.

SALES, W. F., Machado, A. R., Mello, J. D. B. Influence of Cutting Fluid on the Wear of High Speed Steel Drills. In: IV Seminar on Wear. Associaçao Brasileira de Metalurgia e Materiais, Sao Paulo/SP, Ed: Édile, July 1998. p. 641 to 658.

SCANDIFFIO, I. "Uma Contribuiçao ao Estudo do Corte a Seco e ao Corte com Minima quantidade de Lubrificante em Torneamento de Aço", [A Contribution to Dry Cutting and to Minimum Quantity of Fluid in Steel Turning], MSc Dissertation in Mechanical Engineering - UNICAMP, Campinas, SP, 2000.

SHAW, M. C. Heat-Affected Zones in Grinding Steel. Arizona State University, Tempe, AZ, USA. December 16, 1994. Proceedings: of the CIRP vol. 43/1/1994. p.279-282.

SOTO, M. Towards the New Millennium with High-Performance Tools. The World of Machining. Vol. 4, P 10 - 16, April 2000. Publication of the Coromant Division of Sandvik do Brasil.

SOUZA, S. A. Mechanical Testing of Metallic Materials. Theoretical and Practical Foundations. Sao Paulo Ed: Edgard Blucher, 1982, 286p.

TARASOV, L. P. THIBAULT, N. W., Trans. ASM, (38), p331, 1947.

VIEIRA, M. Evaluation of Grinding Wheel Hardness through the Use of Acoustic Emission in Dressing. PhD Thesis, EESC - USP, 1996. P. 131.

VIEIRA, M. J. LIMA, A. LIBARDI, R., CANCILIERI, H. A. Analysis of the Heat Affected Layer on the Surface of Ground Parts. In: IV Seminar on Wear. Associaçao Brasileira de Metalurgia e Materiais, Sao Paulo/SP, Ed: Édile, July 1998.p. 183 to 206.

VILELLA, R. C. Practical Methodology for Optimizing Machining Conditions in Manufacturing Cells. Faculty of Engineering Department of Manufacturing, State University of Campinas, 1998. 104 p, Dissertation (Master's Degree).

WADA, R. There Must Be a Very Different Concept for Tomorrow's Machines. Machines and Metals, Vol. 376, p20 - 43, May 1997. Ed. Aranda.

WEINGAERTNER, W. L. SCHROETER, R. B. Tecnologia de Usinagem das Ligas de Aluminio e Suas Ligas. 2° edição, Sao Paulo: Alcan: Aluminio do Brasil, 1991.

YUHARA, D. A. Drilling with Interchangeable Carbide Inserts. Mundo Mecânico, Vol. 74, p37 - 41, September 1982. Ed. Gruenwald Ltda.

SOUZA, S. A. Technical Testing of Metallic Materials: Theoretical and Practical [illegible]
Paulo: Edgard Blücher, 1982 [illegible]

[illegible]

Printed by Books on Demand GmbH, Norderstedt / Germany